MEDICAL
INTELLIGENCE
UNIT 32

Maspin

Mary J.C. Hendrix, Ph.D., D.Sc.
Department of Anatomy and Cell Biology
Holden Comprehensive Cancer Center
The University of Iowa
Iowa City, Iowa, U.S.A.

CRC Press
Taylor & Francis Group
Boca Raton London New York

CRC Press is an imprint of the
Taylor & Francis Group, an **informa** business

Maspin

Medical Intelligence Unit

Designed by Jesse Kelly-Landes

CRC Press
Taylor & Francis Group
6000 Broken Sound Parkway NW, Suite 300
Boca Raton, FL 33487-2742

First issued in hardback 2019

ISBN 13: 978-1-58706-097-7 (hbk)
ISBN 13: 978-1-58706-098-4 (pbk)

Visit the Taylor & Francis Web site at
http://www.taylorandfrancis.com

and the CRC Press Web site at
http://www.crcpress.com

Library of Congress Cataloging-in-Publication Data

Maspin / [edited by] Mary Hendrix.
p. ; cm. -- (Medical intelligence unit; unit 32)
Includes bibiographical references and index.
ISBN 1-58706-097-3 (hardcover) 1-58706-098-1 (softcover)
1. Antioncogenes.
[DNLM: 1. Serpins--physiology. 2. Gene Expression Regulation. 3. Genes, Suppressor, Tumor--drug effects. 4. Proteins--physiology. QU 136 M412 2001]
I. Hendrix, Mary. II. Series.
RC268.43 .M375 2001
616.99'4061--dc21 2001004699

Dedication

This book is dedicated as a loving tribute to Dr. Ruth Sager, an outstanding geneticist, and her favorite gene—Maspin, mammary serpin protease inhibitor. For more than half a century she demonstrated vision, insight and determination to develop novel scientific concepts in the face of established dogmas. Her pioneering research and original ideas continue to make contributions to biology.

With fond memories and great admiration,
Mary J.C. Hendrix

Acknowledgements

The compilation of manuscripts and final draft of this publication would not have been possible without the hard work and dedication of Shawn Albaugh Kleppe, Department of Anatomy and Cell Biology at The University of Iowa College of Medicine.

CONTENTS

EDITOR

Mary J.C. Hendrix, Ph.D., D.Sc.
Department of Anatomy and Cell Biology
Holden Comprehensive Cancer Center
The University of Iowa
Iowa City, Iowa, U.S.A.
mary-hendrix@uiowa.edu
Chapters 7, 9

CONTRIBUTORS

Mary L. Alpaugh
Department of Pathology
UCLA School of Medicine
Los Angeles, California, U.S.A.
Chapter 4

Sanford H. Barsky
Department of Pathology
UCLA School of Medicine
Los Angeles, California, U.S.A.
sbarsky@ucla.edu
Chapter 4

Hector Biliran Jr.
Wayne State University School of Medicine
Detroit, Michigan, U.S.A.
Chapter 5

Anuja Dokras-Jagasia
Department of Obstetrics/Gynecology
University of Iowa Health Care
Iowa City, Iowa, U.S.A.
anuja-dokras@uiowa.edu
Chapter 9

Frederick E. Domann
Radiation Research Laboratory
The University of Iowa
Iowa City, Iowa, U.S.A.
frederick-domann@uiowa.edu
Chapter 6

Bernard W. Futscher
Radiation Research Laboratory
The University of Iowa
Iowa City, Iowa, U.S.A.
Chapter 6

Lynn M.G. Gardner
Department of Obstetrics/Gynecology
University of Iowa Health Care
Iowa City, Iowa, U.S.A.
Chapter 9

Mickey C-T. Hu
University of Texas
M.D. Anderson Cancer Center
Houston, Texas, U.S.A.
Chapter 10

Mein-Chie Hung
University of Texas
M.D. Anderson Cancer Center
Houston, Texas, U.S.A.
mhung@mail.mdanderson.org
Chapter 10

Paul Kedeshian
Department of Pathology
UCLA School of Medicine
Los Angeles, California, U.S.A.
Chapter 4

Dawn A. Kirschmann
Department of Obstetrics/Gynecology
University of Iowa Health Care
Iowa City, Iowa, U.S.A.
Chapter 9

Richard McGowen
Wayne State University School of Medicine
Detroit, Michigan, U.S.A.
Chapter 5

Valerie A. Odero
Department of Anatomy and Cell Biology
The University of Iowa
Iowa City, Iowa, U.S.A.
Chapter 7

Arthur B. Pardee
Dana Farber Cancer Institute
Boston, Massachusetts, U.S.A.
pardee@mbcrr.harvard.edu
Chapters 1, 2

Phil A. Pemberton
Arriva Pharmaceuticals, Inc.
Alameda, California, U.S.A.
PPemberton@ArrivaPharma.com
Chapter 3

Elisabeth A. Seftor
Department of Anatomy and Cell Biology
The University of Iowa
Iowa City, Iowa, U.S.A.
Chapters 7, 9

Richard E.B. Seftor
Department of Anatomy and Cell Biology
The University of Iowa
Iowa City, Iowa, U.S.A.
richard-seftor@uiowa.edu
Chapter 7

Shijie Sheng
Wayne State University School of Medicine
Detroit, Michigan, U.S.A.
ssheng@med.wayne.edu
Chapters 2, 5

Weiya Xia
University of Texas
M.D. Anderson Cancer Center
Houston, Texas, U.S.A.
Chapter 10

Ming Zhang
Department of Molecular and Cellular Biology
Baylor College of Medicine
Houston, Texas, U.S.A.
Mzhang@bcm.tmc.edu
Chapters 2, 8

PREFACE

This book represents the first compilation of the major research findings regarding maspin—a mammary serpin (or serine) protease inhibitor. Maspin was originaly discovered through subtractive hybridization and differentital display analyses, comparing the differences in gene expression in normal mammary epithelium and invasive carcinoma cells. This work originated in the laboratory of Dr. Ruth Sager, a world-renowned geneticist, at the Dana Farber Cancer Institute and Harvard Medical School. The first introduction of maspin to the research community occurred as a report in *Science* (263:526-529, 1994) and described the unique tumor-suppressing activity of this novel gene in human mammary epithelial cells.

At the time, the data showed that maspin is found in normal mammary epithelial cells and is down-regulated in invasive breast carcinomas. maspin encodes a novel serine protease inhibitor (serpin) with a single 3.0kb mRNA and a protein Mr of 42,000; it contains sequencing homology with members of the serine protease inhibitor superfamily (serpins), including plasminogen activator inhibitor-1, -2 (PAI-1 and PAI-2) and α1-antitrypsin, as well as sequence homology with noninhibitor serpinssuch as ovalbumin. Also, maspin was shown to have tumor suppressor activity when re-expressed in aggressive cancer cells. Furthermore, treatment of breast cancer cells with recombinant maspin inhibited tumor cell motility and invasion in vitro, and this inhibitory action could be reversed with antibodies specific for the reactive loop site. Thus, this original report about maspin offered substantial promise and opportunities with respect to its development as a favorable prognostic marker in breast cancer, in addition to Maspin's potential use in new therapeutic strategies. This publication served as the foundation from which all other maspin studies would be derived—such as the findings reported in this book.

The maspin book provides an overview of Dr. Ruth Sager—the geneticist, and presents a unique perspective of her personal and professional history, contributed by her loving husband, Dr. Arthur B. Pardee. The first chapter describes the innovation, intellect and enthusiasm that Dr. Sager dedicated to her scientific discoveries. Chapter 2 follows with a historical review of maspin research in the Sager laboratory, and describes the underlying concepts of the maspin work from its initial phase to recent developments and findings. Chapter 3 focuses on the structural aspects of maspin with respect to specific biological activities and anticipated developments. The functional insights provided in this chapter are key to our understanding of how to exploit maspin for effective therapeutic application.

Chapter 4 presents the histopathology perspectives of maspin and offers unique insight into the potential use of maspin as a reliable prognostic indicator of disease progression. Chapter 5 focuses on the biochemical assessment of maspin with respect to the complex pericellular plasminogen activation in cell-matrix interactions. An intriguing hypothetical model is presented for the intricate interactions between maspin and plasminogen activator. Chapter 6 follows with new information regarding the genetic and epigenetic regulation of maspin gene expression in normal and tumor tissue, and addresses the mechanisms underlying maspin gene silencing—a viable target for pharmacological reactivation. Chapter 7 describes some interesting observations of the suppression of breast cancer cell invasiveness via Maspin's modulation of integrin expression, and provides additional biological targets for Maspin's activity.

Chapters 8, 9 and 10 are interrelated and complementary. Chapter 8 provides a biological role for maspin in tumor progression and normal development, and highlights a critical role for maspin in angiogenesis. Chapter 9 follows with a unique role for maspin during human placental development. Chapter 10 provides important and relevant data regarding the prognostic potential of maspin in patients with oral squamous cell carcinoma.

The contributors selected for the first maspin book are leaders in their respective fields of research. Several of the contributors were part of the original maspin research with Dr. Ruth Sager. Those of us who had the privilege of knowing and working with Ruth Sager are truly fortunate. She contributed many seminal discoveries to science, and maspin represents a major component of her legacy.

Mary J.C. Hendrix

CHAPTER 1

Ruth Sager, Geneticist

Arthur B. Pardee

Ruth Sager named her favorite gene Maspin, mammary serpin protease inhibitor. Expression of this gene is lost in advanced breast cancers and inhibits tumor invasion and metastasis. This book highlights advances made in studies that developed from her laboratory's researchers starting over a half dozen years ago. Vigorous research on maspin continues in numerous laboratories, including those of several of her past students and fellows.

Very early Ruth believed that genetics is the core of biology. She "knew she was right—and set out to prove it." She has never ceased introducing new techniques and concepts into her field. But she found her work ignored for years—until her discoveries proved the majority wrong. "I don't really pay a lot of attention to what other people think".

Ruth focused her efforts on the genetics of cancer during her final 25 years. "I had really wanted to work on cancer, but it seemed like a very difficult thing to do." Entry into the subject was by her sabbatical at the Imperial Cancer Research Fund in 1972-73. "We think that the first change in cancer is a genetic change — something acts to transform an individual cell—whether that something be a viral infection, or a chemical or radiation".

The first question she asked was which genes cause normal cells to become cancer cells. Growth-stimulating oncogenes, recently discovered, were then being proposed as the basis of cancer. But Sager championed inactivations of tumor suppressor genes, "Nature's own approach to cancer protection", as being in addition deeply involved, and that they are "a vast untapped resource for anticancer therapy". She suggested a Yin-Yang balance of these—motors and brakes of growth—for normal cellular homeostasis.

She investigated changes, between closely related normal and cancer cells, of amounts of specific mRNAs, in particular those whose expression is lost in breast cancers, as underlying the cancer phenotype. She first discovered the IL-8 related GRO gene and others including maspin, whose mRNA levels are down-regulated in tumor cells, by using subtractive hybridization. She then shifted to the then new and simpler differential display technique to discover numerous additional potential tumor suppresssor genes.

Dr. Sager discovered that most of the down-regulated genes are not mutated. But rather their mRNAs are under-expressed, for example by decreased transcription rates. She introduced the concept of "Expression Genetics" now also known as Functional Genomics, the study of changed gene expressions, as compared to genetics which is based upon mutational changes in sequences of structural gene. And she began investigation of methods to reactivate their expression.

Maspin, edited by Mary J.C. Hendrix. ©2002 Eurekah.com

Personality

Ruth Sager was innovative, highly intelligent, enthusiastic, very dedicated to her science, hard working, had high standards, and expected equal dedication from her coworkers. She did not suffer fools gladly. She stated her view of science as a career. "The first thing is to be sure of your own abilities. Science is very demanding, you have to be able to think very well and also have a very good memory. You have to really love it. Science is a way of life. I think it all comes from inside. It really gets to the very core of your existence. It is much like being an artist or a dancer. It's something that demands everything from you that you are capable of." "I have always been intrigued by the physicists approach to scientific inquiry, particularly in the fact that the way to find out something really new is to question the basic tenet of existing theory".

She was described in her fifties as "a calmly articulate and attractive woman (who looks younger by about 15 years)." Again, as "a tall, striking brunette with a ready smile and a voice that carries a merry lilt." She described herself as "probably the happiest person I know."

She was not at all narrowly devoted to her science, and had numerous outside interests—modern art, travel, music and theater, and a rich social life. She was a fine cook. She took up tennis late in life, and played it with great enthusiasm in spite of limited ability. She was especially fond of relaxing at Woods Hole where she had a cherished second home, where she is buried, and where her papers are stored in the Marine Biological Laboratory library.

Personal History

Ruth Sager was born in Chicago on February 7, 1918, daughter of Leon Sager, a businessman with strong intellectual interests, and Deborah Borovik Sager who died in the 1918 influenza epidemic. She and her sisters Esther and Naomi were brought up by her step-mother Hannah in an atmosphere honoring learning. She graduated at 16 from New Trier High School. She received a B.S. in mammalian physiology in 1938 at the University of Chicago, "the best thing that ever happened to me". Her interest in science was originated by Anton J. Carlson's lectures. "He was just a fantastic teacher." She received a M.S. in 1944 in plant physiology at Rutgers Univ., spent the World War II years as a secretary and apple farmer. Her Ph.D. in 1948 was under Marcus M. Rhoades in maize genetics at Columbia University. Then she was a Merck Postdoctoral Fellow 1949-1951 with Sam Granick at the Rockefeller Institute, where she worked on the chloroplast. She was a staff member at the Rockefeller Institute from 1951-1955, where she chose the alga *Chlamydomonas reinhardi* as a model organism. She was a Research Scientist 1955-1965 at Columbia University, and worked for a year in Edinburgh 1962.

For 20 years she could not obtain a faculty position, till age 48. "I guess I knew I was right, and I wasn't terribly upset." She was a Professor at Hunter College 1966-1975, Guggenheim Fellow at Imperial Cancer Research Fund in London 1972-1973. She was finally in 1975 appointed Professor of Cellular Genetics at Harvard Medical School, among the first women to gain Full Professorship at Harvard, and Chief Division of Cancer Genetics, Dana-Farber Cancer Institute.

She was first married to Seymour Melman in 1944, and then to Arthur B. Pardee in 1973. She had no children. She died March 29, 1997, of bladder cancer, in her home in Brookline, Mass. at the age of 79. She worked to the end, publishing innovative articles and obtaining an NIH grant in the month before her death. She is survived by sisters Esther Altschul and Naomi Sager, and her husband.

Professional History

Among her honors and distinctions are Phi Beta Kappa 1938, Sigma Xi 1947, Guggenheim Fellowship in 1972, Schneider Memorial Lecture Award 1973, National Academy of Sciences 1977, American Academy of Arts and Sciences 1978, Harvey Society Lecture 1984, Outstanding Investigator National Cancer Institute 1985, Gilbert Morgan Smith Medal National Academy of Sciences 1988, Institute of Medicine 1992, Princess Takamatsu Award Japan 1992, Alumni Medalist University of Chicago 1994, Advisory Council National Institute on Aging, and other honors.

Cancer Biology

Among her outstanding contributions, Dr. Sager emphasized the major role of chromosome rearrangements and the accelerated evolution of cancer cells, the requirement in a cancer of more than one mutated gene, and importantly of tumor suppressor genes in addition to oncogenes. She proposed as early as 1974 that individual genetic defects could be corrected by transferring DNA into cells. "One need not be doomed by one's genes." She was a pioneer in the novel subject she named expression genetics, the identification by their mRNAs of genes that are functionally modified in cancers. She successfully identified numerous genes that are not mutated but whose expressions are altered in breast cancers, such as the mammary serpin Maspin.

1. She devised the first cell lines and culture medium capable of culturing and comparing normal and cancer cells. She developed a model system that allows detailed comparisons in the same culture medium between well-matched normal and tumor Chinese hamster embryo fibroblasts (CHEF cells).
2. She emphasized the multigenic basis of tumorigenicity. As with her earlier work, she was among the first to champion a then unpopular view, at a time when all attention was on mutations of growth stimulating single oncogenes such as ras. In the late 70's she initiated investigations on negatively regulating genes—tumor suppressors. As with her research on chloroplast genetics, " there was really no interest in tumor suppressor genes at all until about maybe....1990." She demonstrated tumor-suppressing activity with cell hybrids and cybrids. Remarkable examples are suppressor genes such as p53 that promote programmed cell death of defective cells, and these are inactivated in tumors.
3. She provided an initial example of increased genetic instability in cancer cells. Amplification of the methotrexate resistance gene developed much faster in tumor cells than in normal cells.
4. Turning from then popular rodent cells, she decided that gene expression would best be investigated in human cells. For this purpose she created a workable human breast cancer cell culture system, in which epithelial normal and tumor cells could both grow, and at similar rates.
5. Among her final interests were effects of methylation of DNA, and its specific enzymatic cutting, and chromosome rearrangements in tumor cells.

6. The under-expressed genes, including maspin, are not mutated, unlike classical tumor-related genes. Therefore her plan was to use these under-expressed genes as markers for detection and diagnosis, and she hoped to design therapies based on restoring their functions.

Cytoplasmic Inheritance

Ruth Sager's contributions in cancer research were during her second distinguished career, a major innovator in cancer genetics in proposing, discovering, and investigating roles of tumor suppressor genes including maspin. In her first career she was at the pinnacle of research on the problem of non-nuclear or "cytoplasmic" genetics. She almost single handedly developed this subject of non-Mendelian genetics. "A vast, unexplored region of genetics was opened here today (1963)." The very existence of hereditary determinants other than nuclear genes was doubted by a large part of the scientific community, although it was proposed in 1908 from observations on higher plants. Dr. Sager gathered data and argued in support of a second genetic system in the face of great skepticism, and finally made this a respectable and exciting major area of genetics.

At the beginning of her research, Dr. Sager saw the advantages of studying genetics with a model microorganism that had a chloroplast, a sexual life cycle, grew rapidly, and was readily manipulated for controlling growth and mating. She chose the single cell alga "*Chlamydomonas*, a peerless group of organisms. .... nutritious, aesthetically pleasing, and amenable to laboratory experimentation". With talented collaborators, especially her right hand colleague Zenta Ramanis she:

1. Developed a mating system for the organism.
2. Early investigated the genetics of the organism—both Mendelian and non-Mendelian—with clear demonstration of the maternal inheritance pattern of the latter.
3. Discovered, with Y. Tsubo, the first specific "cytoplasmic" gene mutagen, streptomycin, and identified mutants by their resistance to this drug.
4. Discovered ribosomes in the choloroplast of *Chlamydomonas*, different from those in the cytoplasm, thus providing evidence that expression of genetic information as proteins is carried out by a different system.
5. Discovered, with M. R. Ishida, that unique DNA is located in isolated chloroplasts. This was the evidence that convinced most scientists that there is indeed a separate non-nuclear organelle genetic system.
6. Performed biochemical studies of the mechanism of exclusion of paternal genes.
7. Developed a system that makes genetic mapping possible, by permitting the expression of paternal as well as maternal genes.
8. Developed several mapping methods, and first published cytoplasmic linkage groups and extensive mapping of an organelle, showing that the chloroplast's DNA is circular.
9. Demonstrated with an in vitro system the basis of maternally inherited drug resistance.
10. Discovered an eukaryotic restriction enzyme.
11. Discovered that there is communication between nucleus and organelles; they send molecular signals back and forth.
12. Showed that maternal DNA is methylated and paternal DNA is not, and proposed this difference as the basis of selective destructive elimination of the paternal DNA.

Personal Legacy

When asked near the end of her life what she considered to be her most important contribution, she answered "Well, I don't think I've made it yet." Many colleagues have carried on her research, and papers based upon her research continue to appear. She was a major constructive force in the scientific and personal lives of her many friends and students.

She was a role model for many women, being among the earliest very successful woman scientists in spite of major career obstacles. "For more than half a century Ruth Sager has been a role model for women in health-related scientific research. ..... she demonstrated vision, insight and determination to develop novel scientific concepts in the face of established dogmas. .... her pioneering research and original ideas continue to make contributions to biology". But she was never highly active in the women's liberation movement. When faced with the built-in prejudice of the male scientific community against women "there was nothing I could do, except to be as good (a scientist) as possible".

She had great concerns in 1994 about politics and the future of science. "The strong influence of fashions in scientific thought continues to play an inhibitory role in scientific progress. I think science is in a rut right now. The way grants are given out just makes matters worse, because the experiment has to be so obvious and practically done already before they'll fund it." Her career twice demonstrated that some of the best science needs faith and support of novel ideas from the most creative minds.

Articles about Ruth Sager

I am indebted to these authors for many of the cited facts and quotes.

1. Campbell A. The science of persistence. U Chicago Magazine1994; 8:32-35.
2. Ruth Sager. Cover legend. Cancer Res 1996; 56(1).
3. Grinstein LS, Biermann CA, Rose RK, eds. Women in the Biological Sciences. Westport: Greenwood Press, 1997:463-476.
4. Reynolds MD. American Women Scientists—23 Inspiring Biographies 1900-2000. Jefferson: McFarland and Co., 1999:119-127.
5. Bierman CA, Rose RK. Jewish Women in the Sciences. Westport: Greenwood Press, 2001. In press.
6. Schmitt G. Thesis on Ruth Sager. 2001, Princeton, N.J. In preparation.
7. Sapp J. Beyond the Gene: Cytoplasmic Inheritance and the Struggle for Authority in Genetics. New York: Oxford Press, 1987.
8. Ruth Sager, Ph.D. Women in Medicine. Video Cassette, Harvard Medical School, 1982.

Selected Bibliography

Dr. Sager published two ground breaking books:

With Ryan FJ. Cell Heredity. New York: John Wiley and Sons, 1961.
Cytoplasmic Genes and Organelles. New York: Academic Press, 1972.

She published more than 200 articles, among them:

Cancer Research

With Howell N. Tumorigenicity and its suppression in cybrids of mouse and Chinese hamster cell lines. Proc Natl Acad Sci USA 1978; 75:2358-2367.

With Smith BL. The multistep origin of tumor-forming ability of CHEF cells. Cancer Res 1982; 42:389-496.

Genetic suppression of tumor formation: A new frontier in cancer research. Cancer Res 1986; 46:1573-1580.

With O'Brien WG, Stenman G. Suppression of tumor growth by senescence in virally transformed cells. Proc Natl Acad Sci USA 1986; 83:8659-8663.

With Anisowicz A et al. Functional diversity of gro gene expression in human fibroblasts and mammary epithelial cells. Proc Natl Acad Sci USA 1988; 85:9645-9649.

With Band V. Distinctive traits of normal and tumor-derived human mammary epithelial cells expressed in a medium that supports long-term growth of both cell types. Proc Natl Acad Sci USA 1989; 86:1249-1253.

With Kaden D et al. High frequency of large spontaneous deletions of DNA in tumor-derived CHEF cells. Proc Natl Aca Sci USA 1989; 86:2306-2310.

Tumor suppresser genes: The puzzle and the promise. Science 1989; 246:1406-1412.

With Band V et al. A newly established metastatic breast tumor cell line with integrated amplified copies of cerb B-2 and double minute chromosomes. Genes Chromosomes Cancer 1989; 1:48-58.

With Band V et al. Human papilloma virus DNAs immortalize normal human mammary epithelial cells and reduce their growth factor requirements. Proc Natl Acad Sci USA 1990; 87:463-467.

With Zajkowski D et al. Suppression of tumor forming ability and related traits in MCF7 breast cancer cells by fusion with immortal breast epithelial cells. Proc Natl Acad Sci USA 1990; 87:2314-2318.

With Trask D et al. Keratin proteins as markers of normal and tumorigenic breast epithelial cells. Proc Natl Acad Sci USA 1990; 87:2319-2323.

With Swisshelm K et al. Enhanced expression of an insulin growth factor-like binding protein (mac25) in senescent human mammary epithelial cells and induced expression with retinoic acid. Proc Natl Acad Sci USA 1995; 92:4472-4476.

Expression genetics in cancer: Shifting the focus from DNA to RNA. Proc Natl Acad Sci USA 1997; 94:952-955.

With Sheng S, Pemberton P, Hendrix MJ. Maspin: A tumor suppressing serpin. Adv Exp Med Biol 1997; 425:77-88.

With Martin KJ et al. Linking gene expression patterns to therapeutic groups in breast cancer. Cancer Res 2000; 60:2232-2238.

Cytoplasmic Inheritance

With Granick S. Nutritional control of sexuality in *Chlamyodomonas reinhardi*. J Gen Physiol 1954; 37:729-742.

Non-Mendelian inheritance of streptomycin resistance in *Chlamyodomonas reinhardi*. Proc Natl Acad Sci USA 1954; 40:356-363.

Inheritance in the green alga *Chlamyodomonas reinhardi*. Genetics 1955; 40:476-489.

Streptomycin as a mutagen for nonchromosomal genes. Proc Natl Acad Sci USA 1962; 48:2018.

With Ishida MR. Chloroplast DNA in *Chlamydomonas*. Proc Natl Acad Sci USA 1963; 50:725-730.

With Ramanis Z. Recombination of non-chromosomal genes in *Chlamydomonas*. Proc Natl Acad Sci USA 1965; 53:1053-1061.

With Hamilton MG. Cytoplasmic and chloroplast ribosomes of *Chlamydomonas*. Ultracentrifugal characterization. Science 1967; 157:709-711.

With Ramanis Z. Biparental inheritance of non-chromosomal genes induced by ultraviolet irradiation. Proc Natl Acad Sci USA 1967; 58:931-937.

With Ramanis Z. A genetic map of non-Mendelian genes in *Chlamydomonas*. Proc Natl Acad Sci USA 1970; 65:593-600.

With Lane D. Molecular basis of maternal inheritance. Proc Natl Acad Sci USA 1972; 69:2410-2413.
With Roberts RJ, Myers PA. A site-specific single strand endonuclease from the eukaryote *Chlamydomonas*. Proc Natl Acad Sci USA 1977; 74:2687-2691.
With Burton WG, Grabowy C. The role of methylation in the modification and restriction of chloroloplast DNA in *Chlamydomonas*. Proc Natl Acad Sci USA 1979; 76:1390-1394.
With Sano H, Grabowy C. Differential activity of DNA methyltransferase in the life cycle of *Chlamyodomonas reinhardi*. Proc Natl Acad Sci USA 1981; 78:3118-3122.

Chapter 2

Maspin in the Sager Laboratory

Ming Zhang, Shijie Sheng and Arthur B. Pardee

Introduction and Underlying Concepts

Discovery of a disease-related gene marks only the beginning to a series of difficult investigations. In order to establish the functional role of the newly discovered gene, one has to obtain insights into its biological activities, genetic and epigenetic regulations, and molecular mechanisms operating under various conditions such as in cancer cells and in normal cells. Is it, as examples, involved in growth control, genetic stability, apoptosis, immortalization, differentiation, angiogenesis, invasion and metastasis, etc? What is the molecular biology of its transcription? Does it in turn control the transcription of other genes? Where is its protein product located? What are the biochemical properties of the protein? Is it an enzyme, transcription factor or a structural element, etc? Does it undergo post-translational modifications, and what effects do these have?

The following is a summarization of published work on maspin performed by researchers in and from the Ruth Sager laboratory. In order to preserve the flavor of the laboratory, the material that follows is composed of the summaries of their publications. It is intended to provide a general paradigm of how the discovery of one gene of interest can lead into a subject for intensive investigation and to novel concepts. And also it illustrates approaches that can be applied to establish functional roles of a newly discovered gene.

Tumor Suppressors

Tumor Suppressor Genes: The Puzzle and the Promise[1]

Tumor suppressor genes are wild-type alleles of genes that play regulatory roles in cell proliferation, differentiation, and other cellular and systemic processes. It is their loss or inactivation that is oncogenic. The first evidence of tumor suppressor genes appeared in the early 1970s, but only within the past few years has a wealth of new information illuminated the central importance of these genes. Several different suppressor genes may be inactivated in the same tumors, and the same suppressors may be inactive in different tumor types (for example, lung, breast, and colon). The suppressor genes already identified are involved in cell cycle control, signal transduction, apoptosis, angiogenesis, and development, demonstrating that they contribute to a

Maspin, edited by Mary J.C. Hendrix. ©2002 Eurekah.com

broad array of normal and tumor-related functions. It is proposed that tumor suppressor genes provide a vast untapped resource for anticancer therapy.

Expression Genetics (Subtractive Hybridization and DD)

Expression Genetics in Cancer: Shifting the Focus from DNA to RNA[10]

Expression genetics is a conceptually different approach to the identification of cancer-related genes than the search for mutations at the genome level. While mutations lie at the heart of cancer, at least in its early stages, what is recognized here are phenotypic changes that are usually many steps removed from the initiating mutation. Cancer geneticists have classically concentrated on genomic changes and have ignored the productive potential of examining downstream events based on screening for differential gene expression between tumor cells and well matched normal counterparts. Genes involved in cancer affect the normal functions of many cellular processes, not only proliferation but cell-cell and cell-matrix interactions, DNA repair, invasion and motility, angiogenesis, senescence, apoptosis, and others. However, very few cancer-related genes affecting these processes have been identified in human cancers by classical methods despite enormous efforts. I report here our success in readily isolating more than 100 candidate tumor suppressor genes from human tissue, estimated to represent roughly 20% of the total genes recoverable by this approach. Half of the genes are unknown and the other half includes representatives of most known cancer processes. Because their expression is lost during cancer progression, they may be useful tumor markers for diagnosis and prognosis. Because several of these genes are not mutated, they provide opportunities for pharmacological intervention by inducing their re-expression.

Expression Genetics: A Different Approach to Cancer Diagnosis and Prognosis[14]

Expression genetics is a new approach to the identification of cancer-related genes. Instead of studying mutations at the genome level (gene mutations), expression genetics is the investigation of heredity at the RNA level. By isolating genes whose expression is up- or down-regulated in cancers, expression geneticists study their function in the context of gene regulation. A major goal of expression genetics in cancer is to correct gene expression in tumors by the application of potential therapeutic agents.

The Biology of Human Maspin

Identification of Maspin

RNA Genetics of Breast Cancer: Maspin as a Paradigm[4]

Our focus is on genes that are down-regulated, but not mutated, in mammary carcinomas but not in normal mammary epithelial cells. This focus has led to the identification of a large number of candidate tumor suppressor genes, approaching the number of known oncogenes. This research is the initial demonstration that dysregulated genes are numerous in cancer. Implications of the fact that so many genes are simultaneously down-regulated but not mutated in cancer cells are considered. Expansions of such gene sets are currently under vigorous investigation, by applying powerful methods including Differential Display, SAGE, and microarrays.

Maspin is an example of a down-regulated gene in breast cancer, discovered in this laboratory by subtractive hybridization. Maspin encodes a novel serine protease inhibitor (serpin), as demonstrated by sequence comparison. Maspin can be re-expressed in tumor cells by phorbol ester treatment. It functions as a tumor suppressor gene.

Maspin: A Tumor Suppressing Serpin[6]

Maspin, a serpin found in normal mammary epithelial cells, is down-regulated in invasive breast carcinomas. Similar patterns of expression at the RNA and protein levels are seen by Northern analysis with cells grown in culture and by immunostaining of tissues. Maspin has been shown to have tumor suppressor activity. Although maspin does not behave as a classical inhibitory serpin against any known target protease in solution. Biological studies have shown that recombinant maspin inhibits tumor cell motility and invasion through reconstituted basement membranes, and that its inhibitory action is totally lost by a single cleavage at the reactive loop site. Tumor transfectants expressing maspin are inhibited in growth and metastasis in nude mice. The biological function of maspin is located at the cell surface.

Maspin, a Serpin with Tumor-Suppressing Activity in Human Mammary Epithelial Cells[2]

A gene encoding a protein related to the serpin family of protease inhibitors was identified as a candidate tumor suppressor gene that may play a role in human breast cancer. The gene product, called maspin, is expressed in normal mammary epithelial cells but not in most mammary carcinoma cell lines. Transfection of MDA-MB-435 mammary carcinoma cells with the maspin gene did not alter the cells' growth properties in vitro, but reduced the cells' ability to grow and metastasize in nude mice and to invade through a basement membrane matrix in vitro. Analysis of human breast cancer specimens revealed that loss of maspin expression occurred most frequently in advanced cancers. These results support the hypothesis that maspin functions as a tumor suppressor. For a review see ref. 12.

In Breast Cancer vs. Normal Cells

Production, Purification and Characterization of Recombinant Maspin Proteins[3]

In this paper, we report the production of recombinant glutathione S-transferase-maspin fusion protein, expressed in the bacterium *Escherichia coli,* and recombinant maspin, expressed in the insect *Spodoptera frugiperda* cells. The GST-fusion protein was purified by glutathione affinity chromatography. Maspin expressed in insect cells was purified by a combination of Bio-Rad AG1-2X anion exchange chromatography and heparin affinity chromatography. Both recombinant proteins demonstrated strong inhibitory effects on the invasion by two breast tumor cell lines across reconstituted basement membranes and such inhibitory effect was abolished in the presence of the polyclonal antibody made against the reactive center region of maspin. The recombinant maspin from insect cells was cleaved by trypsin specifically at the putative reactive center, as confirmed by protein sequencing. The trypsin-cleaved recombinant maspin did not inhibit invasion, indicating that intact putative reactive center of maspin is required for its biological activities. This paper provides evidence that recombinant maspin protein itself inhibits invasion and supports the role of maspin as a tumor suppressor.

Maspin Acts at the Cell Membrane to Inhibit Invasion and Motility of Mammary and Prostatic Cancer Cells[7]

Recombinant maspin protein blocks the motility of breast carcinoma cells in culture over 12 h, as demonstrated by time-lapse video microscopy. Lamellopodia are withdrawn but ruffling continues. Both exogenous recombinant maspin and maspin expressed by tumor transfectants exhibit inhibitory effects on cell motility and cell invasion as shown in modified Boyden chamber assays. When mammary carcinoma cells were treated with recombinant maspin, the protein was shown by immunostaining to bind specifically to the cell surface, suggesting that maspin activity is membrane associated. When pretreated with antimaspin antibody, maspin loses its inhibitory effects on both invasion and motility. However, when maspin is added to these cells preceding antibody treatment, the activity of maspin is no longer inhibited by subsequent addition of the antibody. It is concluded therefore that the inhibition of invasion and motility by maspin is initially localized to the cell surface.

Maspin Suppresses the Invasive Phenotype of Human Breast Carcinoma[15]

The exploitation of maspin as a potential diagnostic and/or therapeutic tool has remained limited due to the lack of knowledge concerning its molecular and biological mechanism(s) of action. The work reported here demonstrates that treatment of MDA-MB-435 cells with recombinant maspin enhances the selective cell adhesion to a fibronectin matrix and induces the conversion from a fibroblastic to a more epithelial-like phenotype. In addition, a blocking antibody to integrins is shown to abrogate the inhibitory activity of maspin on tumor cell invasion through a fibronectin matrix-containing barrier in vitro. Further molecular analyses revealed that recombinant maspin induces higher cell surface levels of some integrins and reduced levels of others in the metastatic human breast carcinoma cell line MDA-MB-435, concomitant with its ability to inhibit the invasive process in vitro. Taken together, these data address the hypothesis that maspin reduces the invasive phenotype of MDA-MB-435 cells by altering their integrin profile, which in turn converts these cells to a more benign epithelial phenotype, with a reduced invasive potential. These data provide new insights into the biological significance of this tumor suppressor gene found in normal mammary epithelium and may form the basis of novel therapeutic strategies in the management of breast carcinoma.

In Prostate Cancer

Maspin Acts at the Cell Membrane to Inhibit Invasion and Motility of Mammary and Prostatic Cancer Cells[7]

Since breast and prostate are both glandular epithelial tissues, prostate cancer and breast cancers have numerous common pathological features. We showed that recombinant maspin exerts a potent inhibitory effect on three prostatic cancer cell lines in both invasion and motility assays. An oligopeptide-derived polyclonal antibody against the reactive site loop sequence of maspin neutralized the inhibitory activity of maspin in both assays. This data suggests that maspin may block the progression of breast cancer and prostate cancer by a similar mechanism. (Additional results with prostate cancer are presented below.)

Mouse Maspin

In Mouse Breast Cancer

mMaspin: The Mouse Homolog of a Human Tumor Suppressor gene Inhibits Mammary Tumor Invasion and Motility[8]

In order to examine the role of maspin in an intact mammal, we cloned mouse maspin cDNA by screening a mouse mammary gland cDNA library with the human maspin cDNA probe. The deduced protein sequence of mouse maspin (mMaspin) is 89% homologous with human maspin. Like its human homolog, mMaspin is expressed in normal mouse mammary epithelial cells and down-regulated in mouse breast tumor cell lines. The expression is altered at different developmental stages in the mammary gland.

The recombinant mouse maspin was produced as a GST-fusion protein in *E. coli* and was purified by glutathione affinity chromatography. Addition of the recombinant mMaspin protein to mouse tumor cells was shown to inhibit invasion in a dose-dependent manner. As with the human protein, recombinant mMaspin protein also inhibited mouse mammary tumor motility. Deletion in the putative mMaspin reactive site loop (RSL) region resulted in the loss of its inhibitory functions. Taken together, these data suggest that the homologous proteins play similar physiological roles in vivo.

Reduced Mammary Tumor Progression in WAP-TAg/WAP-Maspin Bitransgenic Mice[19]

Maspin is a unique serpin involved in the suppression of tumor growth and metastasis. To investigate whether increased levels of maspin protect against tumor progression in vivo, we established a transgenic model in which maspin is targeted to mammary epithelial cells by the whey acidic protein (WAP) promoter for over expression. We crossed these maspin transgenic mice with the WAP-TAg mouse model of tumor progression. Maspin over expression increased the rate of apoptosis of both preneoplastic and carcinomatous mammary epithelial cells. Maspin reduced tumor growth through a combination of reduced angiogenesis and increased apoptosis. The number of pulmonary metastases was reduced with maspin over expression. These data demonstrate that targeted over expression of maspin can inhibit tumor progression in vivo, likely through a combination of increased apoptosis, decreased angiogenesis, and inhibition of tumor cell migration.

In Normal Breast Development

Maspin Plays an Important Role in Mammary Gland Development[16]

Maspin, a unique member of the serpin family, functions as a class II tumor suppressor gene. Despite its known activity against tumor invasion and motility, little is known about maspin's functions in normal mammary gland development. In this paper, we show that maspin does not act as a tissue plasminogen activator inhibitor in the mammary gland. However, targeted expression of maspin by the whey acidic protein gene promoter inhibits the development of lobular-alveolar structures during pregnancy and disrupts mammary gland differentiation. Apoptosis was increased in alveolar cells from transgenic mammary glands at midpregnancy. However, the rate of proliferation was increased in early lactating glands to compensate for the retarded development during pregnancy. These findings demonstrate that maspin plays an important role in mammary development and that its effect is stage dependent.

Biochemistry and Molecular Biology

Transcription Control

Transactivation through ets and Ap1 Transcription Sites Determines the Expression of the Tumor-Suppressing Gene Maspin[9]

Tumor invasion and metastasis are processes poorly understood at the molecular level. Maspin is a serine protease inhibitor (serpin) with tumor-suppressing function in the mammary gland. Maspin gene expression is decreased with malignancy and is lost in metastatic cells. We show in this report that differential expression of maspin in normal and carcinoma-derived mammary epithelial cells is regulated at the transcriptional level. We have identified the ets and Ap1 sites in the maspin promoter that are active in regulating maspin expression in normal mammary epithelial cells but inactive in tumor cells. The ets site alone is sufficient to activate transcription in a heterologous promoter, whereas the Ap1 site cooperates with ets in activation. The enhancing function by ets and Ap1 elements is decreased in primary tumor cells (21NT) and is abolished in invasive tumor cells (MDA-MB-231). Thus, loss of maspin expression during tumor progression results at least in part from the absence of transactivation through the ets and Ap1 sites.

Expression of Maspin in Prostate Cells is Regulated by a Positive ets Element and a Negative Hormonal Responsive Element Site Recognized by Androgen Receptor[11]

Prostate cancer is the most common cancer in men. The molecular mechanisms leading to its development are poorly understood. Maspin is a tumor-suppressing serpin expressed in normal breast and prostate epithelium. We have found that expression of maspin in normal and carcinoma-derived prostate epithelial cells is differentially regulated at the transcriptional level. We have identified two different kinds of cis elements, ets and hormonal responsive element (HRE), in the maspin promoter. The ets element is active in regulating maspin expression in normal prostate epithelial cells but inactive in tumor cells. The HRE site is a negative element that is active in both cell types. This negative DNA sequence can repress a heterologous promoter recognized by the androgen receptor. We conclude that expression of maspin is under the influence of both a positive ets and a negative HRE element. Loss of maspin expression during tumor progression apparently results from both the absence of transactivation through the ets element and the presence of transcription repression through the negative HRE element recognized by androgen receptor.

p53 Regulates the Expression of the Tumor Suppressor Gene Maspin[21]

Maspin has been shown to inhibit tumor cell invasion and metastasis in breast tumor cells. Maspin expression was detected in normal breast and prostate epithelial cells, whereas tumor cells exhibited reduced or no expression. However, the regulatory mechanism of maspin expression remains unknown. We report here a rapid and robust induction of maspin expression in prostate cancer cells (LNCaP, DU145 and PC3) and breast tumor cells (MCF7) following wild type p53 expression from an adenovirus p53 expression vector (AdWTp53). p53 activates the maspin promoter by binding directly to the p53 consensus-binding site present in the maspin promoter. DNA-damaging agents and cytotoxic drugs induced endogenous maspin expression in cells containing the wild type p53. Maspin expression was refractory to the DNA-damaging agents in cells containing mutant p53. These results, combined with recent studies of the tumor metastasis suppressor gene KAI1 and plasminogen activator inhibitor 1 (PAI1), define a new category of molecular targets of p53 that have the potential to negatively regulate tumor invasion and/or metastasis.

Protease Inhibition-Structure and Kinetics

Maspin is a protein of 42 kilodalton with extensive sequence homology to members of the serpin family. Serpins are divided into two classes (inhibitory and non-inhibitory) and both play very important roles in vivo. The following is evidence that supports maspin's role as a protease substrate or inhibitor.

The Tumor Suppressor Maspin Does Not Undergo the Stressed to Relaxed Transition or Inhibits Trypsin-Like Serine Proteases: Evidence that Maspin Is Not a Protease Inhibitory Serpin[5]

The role of tumor suppressor proteins in the development of malignancy has made the understanding of their molecular mechanisms of action of great importance. Maspin is a tumor suppressor produced by a number of cell types of epithelial origin. Exogenous recombinant maspin has been shown to block the growth, motility, and invasiveness of breast tumor cell lines in vitro and in vivo. Although it belongs to the serine proteinase inhibitor (serpin) superfamily of proteins, the molecular mechanism of maspin is currently unknown. Here we show that the reactive site loop of maspin exists in an exposed conformation. The reactive site loop of maspin, however, does not act as an inhibitor of proteinases such as chymotrypsin, elastase, plasmin, thrombin, and trypsin but rather as a substrate. Maspin is also unable to inhibit tissue and urokinase type plasminogen activators. Stability studies show that maspin cannot undergo the stressed-relaxed transition typical of proteinase-inhibitory serpins, and the protein is capable of spontaneous polymerization induced by changes in pH. It is likely, therefore, that maspin is structurally more closely related to ovalbumin and angiotensinogen, and its tumor suppressor activity is independent of a latent or intrinsic trypsin-like serine proteinase-inhibitory activity.

Tissue-Type Plasminogen Activator is a Target of the Tumor Suppressor Gene Maspin[13]

Maspin is structurally a member of the serpin (serine protease inhibitors) superfamily but deviates somewhat from classical serpins. We find that single-chain tissue plasminogen activator (sctPA) specifically interacts with the maspin reactive site loop peptide and forms a stable complex with recombinant maspin. Major effects of maspin are observed on plasminogen activation by sctPA. First, maspin activates free sctPA. Second, it inhibits sctPA preactivated by poly-D-lysine. Third, maspin exerts a biphasic effect on the activity of sctPA preactivated by fibrinogen/gelatin, acting as a competitive inhibitor at low concentrations (< 0.5 micromolar) and as a stimulator at higher concentrations. Fourth, 38-kDa C-terminal truncated maspin further stimulates fibrinogen/gelatin-associated sctPA. Maspin acts specifically; it does not inhibit urokinase-type plasminogen activator, plasmin, chymotrypsin, trypsin, or elastase in solution. Our kinetic data are quantitatively consistent with a model in which two segregated domains of maspin interact with the catalytic and activating domains of sctPA. These complex interactions between maspin and sctPA in vitro suggest a mechanism by which maspin regulates plasminogen activation by sctPA bound to the epithelial cell surface.

The Surface of Prostate Carcinoma DU145 Cells Mediates the Inhibition of Urokinase-Type Plasminogen Activator by Maspin[18]

Maspin is a novel serine protease inhibitor (serpin) with tumor suppressive potential in breast and prostate cancer, acting at the level of tumor invasion and metastasis. It was subsequently demonstrated that maspin inhibits tumor invasion, at least in part, by inhibiting cell motility. Interestingly, in cell-free solutions, maspin does not inhibit several serine proteases including tissue-type plasminogen activator and urokinase-type plasminogen activator (uPA). Despite the recent biochemical evidence that maspin specifically inhibits tissue-type plasminogen activator that is associated with fibrinogen or poly-lysine, the molecular mechanism underlying the tumor-suppressive effect of maspin remains elusive. The goal of this study was to investigate the effect of maspin on cell surface-associated uPA. In our experimental system, we chose prostate carcinoma DU145 cells because these cells mediate plasminogen activation primarily by uPA, as shown by two different colorimetric enzyme activity assays. Purified recombinant maspin produced in baculovirus-infected *Spodoptera frugiperda Sf9* insect cells binds specifically to the surface of DU145 cells, inhibits the DU145 cell surface-bound uPA, and forms a stable complex with the uPA in DU145 cell lysate. The inhibitory effect of maspin on cell surface-bound uPA was similar to that of an uPA-neutralizing antibody and was reversed by a polyclonal antibody against the reactive site loop sequence of maspin. The Ki value for maspin in cell surface-mediated plasminogen activation was 20 nM, which was comparable to the Ki values for plasminogen activator inhibitor 1 and plasminogen activator inhibitor 2, respectively. Furthermore, the proteolytic inhibitory effect of maspin was quantitatively consistent with its inhibitory effect on the motility of DU145 cells in vitro. Our data demonstrate an important role for the prostate carcinoma cell surface in mediating the inhibitory interaction between maspin and uPA. Thus, future maspin-based therapeutic strategies may prove useful in blocking the invasion and metastasis of uPA-positive prostate carcinoma.

Recent Developments

Cancer Marker Detection

Linking Gene Expression Patterns to Therapeutic Groups in Breast Cancer[17]

A major objective of current cancer research is to develop a detailed molecular characterization of tumor cells and tissues that is linked to clinical information. Toward this end, we have identified approximately one-quarter of all genes that were aberrantly expressed in a breast cancer cell line using differential display. The cancer cells lost the expression of many genes involved in cell adhesion, communication, and maintenance of cell shape, while they gained the expression of many synthetic and metabolic enzymes important for cell proliferation. High-density, membrane-based hybridization arrays were used to study mRNA expression patterns of these genes in cultured cells and archived tumor tissue. Cluster analysis was then used to identify groups of genes, the expression patterns of which correlated with clinical information. Two clusters of genes, represented by p53 and maspin, had expression patterns that strongly associated with estrogen receptor status. A third cluster that included HSP-90 tended to be associated with clinical tumor stage, whereas a forth cluster that included keratin 14 tended to be associated with tumor size. Expression levels of these clinically relevant gene clusters allowed breast tumors to be grouped into distinct categories. Gene expression fingerprints that include these four gene clusters have the potential to improve prognostic accuracy and therapeutic outcomes for breast cancer patients.

Detection in Blood

High-Sensitivity Array Analysis of Gene Expression For the Early Detection of Disseminated Breast Tumor Cells in Peripheral Blood[22]

Early detection is an effective means of reducing cancer mortality. Here, we describe a highly sensitive high-throughput screen that can identify panels of markers for the early detection of solid tumor cells disseminated in peripheral blood. The method is a two-step combination of differential display and high-sensitivity cDNA arrays. In a primary screen, differential display identified 170 candidate marker genes differentially expressed between breast tumor cells and normal breast epithelial cells. In a secondary screen, high-sensitivity arrays assessed expression levels of these genes in 48 blood samples, 22 from healthy volunteers and 26 from breast cancer patients. Cluster analysis identified a group of 12 genes that were elevated in the blood of cancer patients. Permutation analysis of individual genes defined five core genes (P </ = 0.05, permax test). As a group, the 12 genes generally distinguished accurately between healthy volunteers and patients with breast cancer. Maspin was an excellent marker in this set. Mean expression levels of the 12 genes were elevated in 77% (10 of 13) untreated invasive cancer patients, whereas cluster analysis correctly classified volunteers and patients (P = 0.0022, Fisher's exact test). Quantitative real-time PCR confirmed array results and indicated that the sensitivity of the assay (1:2 x 10(8) transcripts) was sufficient to detect disseminated solid tumor cells in blood. Expression-based blood assays developed with the screening approach described here have the potential to detect and classify solid tumor cells originating from virtually any primary site in the body.

Therapy

Maspin Is an Angiogenesis Inhibitor[20]

Maspin, a unique member of the serpin family, is a secreted protein encoded by a class II tumor suppressor gene whose downregulation is associated with the development of breast and prostate cancers. Overexpression of maspin in breast tumor cells limits their growth and metastases in vivo. In this report we demonstrate that maspin is an effective inhibitor of angiogenesis. In vitro, it acted directly on cultured endothelial cells to stop their migration towards basic fibroblast growth factor and vascular endothelial and to limit mitogenesis and tube formation. In vivo, it blocked neovascularization in the rat cornea pocket model. Maspin derivatives mutated in the serpin reactive site lost their ability to inhibit the migration of fibroblasts, keratinocytes, and breast cancer cells but were still able to block angiogenesis in vitro and in vivo. When maspin was delivered locally to human prostate tumor cells in a xenograft mouse model, it blocked tumor growth and dramatically reduced the density of tumor-associated microvessels. These data suggest that the tumor suppressor activity of maspin may depend in large part on its ability to inhibit angiogenesis and raise the possibility that maspin and similar serpins may be excellent leads for the development of drugs that modulate angiogenesis.

Exciting opportunities for therapy could result from a network of coordinated regulation in which reversing the down-regulation of one gene could also repress others in a coordinate manner, leading to substantial normalization of the phenotype. Focusing on mechanisms of phenotype reversion guided by reexpression of known down-regulated genes offers substantial hope for the success of nontoxic cancer chemotherapies.[4]

References

1. Sager R. Tumor suppressor genes: The puzzle and the promise. Science 1989; 246:1406-1412.
2. Zou Z, Anisowicz A, Hendrix MJ, Thor A, Neveu M, Sheng S et al. Maspin, a serpin with tumor-suppressing activity in human mammary epithelial cells. Science 1994; 263:526-529.
3. Sheng S, Pemberton PA, Sager R. Production, purification, and characterization of recombinant maspin proteins. J Biol Chem. 1994; 269:30988-30993.
4. Sager R, Sheng S, Anisowicz A, Sotiropoulou G, Zou Z, Stenman G et al. RNA genetics of breast cancer: Maspin as paradigm. Cold Spring Harb Symp Quant Biol 1994; 59:537-546.
5. Pemberton PA, Wong DT, Gibson HL, Kiefer MC, Fitzpatrick PA, Sager R et al. The tumor suppressor maspin does not undergo the stressed to relaxed transition or inhibit trypsin-like serine proteases. Evidence that maspin is not a protease inhibitory serpin. J Biol Chem 1995; 270:15832-15837.
6. Sager R, Sheng S, Pemberton P, Hendrix MJ. Maspin: A tumor suppressing serpin. Curr Top Microbiol Immunol 1996; 213(Pt 1):51-64.
7. Sheng S, Carey J, Seftor EA, Dias L, Hendrix MJ, Sager R. Maspin acts at the cell membrane to inhibit invasion and motility of mammary and prostatic cancer cells. Proc Natl Acad Sci USA 1996; 93:11669-11674.
8. Zhang M, Sheng S, Maass N, Sager R. mMaspin: The mouse homolog of a human tumor suppressor gene inhibits mammary tumor invasion and motility. Mol Med 1997; 3:49-59.
9. Zhang M, Maass N, Magit D, Sager R. Transactivation through ets and Ap1 transcription sites determines the expression of the tumor-suppressing gene maspin. Cell growth Differ 1997; 8:179-186.
10. Sager R. Expression genetics in cancer: shifting the focus from Maspin in the Sager LaboratoryDNA to RNA. Proc Natl Acad Sci USA 1997; 94:952-955.
11. Zhang M, Magit D, Sager R. Expression of maspin in prostate cells is regulated by a positive ets element and a negative hormonal responsive element site recognized by androgen receptor. Proc Natl Acad Sci USA 1997; 94:5673-5678.
12. Sager R, Sheng S, Pemberton P, Hendrix MJ. Maspin. A tumor suppressing serpin. Adv Exp Med Biol 1997; 425:77-88.
13. Sheng S, Truong B, Fredrickson D, Wu R, Pardee AB, Sager R. Tissue-type plasminogen activator is a target of the tumor suppressor gene maspin. Proc Natl Acad Sci USA 1998; 95(2):499-504.
14. Zhang M, Martin KJ, Sheng S, Sager R. Expression genetics: A different approach to cancer diagnosis and prognosis. Trends Biotechnol 1998; 16:66-71.
15. Seftor RE, Seftor EA, Sheng S, Pemberton PA, Sager R, Hendrix MJ. Maspin suppresses the invasive phenotype of human breast carcinoma. Cancer Res 1998; 58:5681-5685.
16. Zhang M, Magit D, Botteri F, Shi HY, He K, Li M et al. Maspin plays an important role in mammary gland development. Dev Biol 1999; 215:278-287.
17. Martin KJ, Kritzman BM, Price LM, Koh B, Kwan CP, Zhang X et al. Linking gene expression patterns to therapeutic groups in breast cancer. Cancer Res 2000; 60:2232-2238.
18. McGowen R, Biliran H Jr, Sager R, Sheng S. The surface of prostate carcinoma DU145 cells mediates the inhibition of urokinase-type plasminogen activator by maspin. Cancer Res 2000; 60:4771-4778.

19. Zhang M, Shi Y, Magit D, Furth PA, Sager R. Reduced mammary tumor progression in WAP-TAg/WAP-maspin bitransgenic mice. Oncogene 2000; 19:6053-6058.
20. Zhang M, Volpert O, Shi YH, Bouck N. Maspin is an angiogenesis inhibitor. Nature Med 2000; 6:196-199.
21. Zou Z, Gao C, Nagaich AK, Connell T, Saito S, Moul JW et al. p53 regulates the expression of the tumor suppressor gene maspin. J Biol Chem 2000; 275:6051-6054.
22. Martin KJ, Graner E, Li Y, Price LM, Kritzman BM, Fournier MV et al. High-sensitivity array analysis of gene expression for the early detection of disseminated breast tumor cells in peripheral blood. Proc Natl Acad Sci USA 2001; 98:2646-2651.

CHAPTER 3

Maspin: Functional Insights from a Structural Perspective

Philip A. Pemberton

Introduction

Since the seminal paper by Zou et al[1] identifying the existence of the novel tumor suppressor maspin (mammary serpin), research efforts have largely focused on the mechanism of action of the protein and its utility as a prognostic indicator for other types of cancer. Maspin regulates tumor cell motility, invasion, growth, and apoptosis.[2,3,4] It has also been identified as a regulator of angiogenesis.[5,40] Recently, several molecular targets for maspin have been discovered which include the integrins and, more controversially, the plasminogen activators uPA and tPA.[6,7,8] As the number of potential targets and our knowledge regarding the regulation of maspin function increases it would be useful to map these binding sites/functions to an overall framework of maspin structure. Despite the absence of a crystallographic structure for maspin, a model structure has been created which provides a useful starting point for this analysis[33]—particularly in light of the wealth of published information on other serpin superfamily members, many of which exhibit similar functions.

Biological Activities of Maspin

Maspin is a Class II Tumor Suppressor Molecule

Maspin was discovered by screening for genes that are down-regulated during the progression from a normal cell phenotype to a tumorigenic, invasive, and metastatic phenotype.[9] Such "Class II" tumor suppressor genes are unlike Class I tumor suppressor genes (e.g., p53, pRb) because they are not expected to harbor any mutations within their coding sequences. Rather, as in the case of maspin, other events are responsible for a down-regulation of gene transcription (e.g., aberrant cytosine methylation, chromatin condensation, Ref. 10). This means that local perturbations in protein structure/function are unlikely to contribute to the complexity of mutational phenotypes that exist for other diseases (e.g., lysosomal enzyme storage diseases[11] and, mutations in serpins that predispose to other diseases, Ref. 12). Thus, reactivation of the gene(s) in the target cell population or, replacement of the protein(s), present as

Maspin, edited by Mary J.C. Hendrix. ©2002 Eurekah.com

viable therapy options in cancer patients. Indeed, as in the case of maspin, recent publications suggest that some currently used chemotherapeutic agents function to do precisely this (e.g., tamoxifen).[13]

Maspin Inhibits Tumor Cell Invasion and Motility

Recombinant maspin protein(s) added to tumor cells in vitro block their ability to invade through artificial matrices that mimic basement membranes.[14,15] The sources of maspin tested to date are a) recombinant proteins with the mature human sequence expressed in, and purified from, insect cells (*S. frugiperda*) and yeast (*S. cerevisciae*) and b) GST-maspin fusion proteins comprising both mouse and human maspin sequences produced in *E.coli*. Interestingly, all these molecules inhibit tumor cell invasion at optimal doses of 100-200 nM but at higher doses the effect of the mature human sequences is lost whereas the GST-human maspin construct maintains its inhibitory activity. These data suggest there to be structural (and functional) differences between the natural forms of the protein and GST-maspin constructs (a potential mechanism of inhibition involving the plasminogen activators uPA and tPA is discussed elsewhere in this book and will not be detailed here). A similar inhibition of invasion has been observed for tumor cells transfected with, and expressing, the maspin gene. Efforts to detect maspin in the media of transfectant cell cultures have proven unsuccessful in this laboratory but it has been reported that exogenous maspin binds to mammary and prostate tumor cell surfaces so it is possible that transfectants secrete maspin but localize it to the cell surface in a similar fashion.[15] In contrast, normal human mammary epithelial cells (HMEC), normal human prostate epithelial cells (HPEC), and normal squamous cells sequester the majority of maspin in the cytoplasm suggesting that part of maspin's tumor suppressing activity may be dependent upon an intracellular function.[16,41] Exogenous maspin also inhibits motility—suggesting that signaling events involved in reorganization of the major components of the cytoskeleton—actin and tubulin are affected. Recent data demonstrating alterations in integrin profiles of tumor cells treated with exogenous is likely one facet of this pathway.[6] Given the nature of the changes induced by maspin, other facets of this pathway will likely involve one or more of the intracellular signaling proteins Rac, Rho, or Ras.[17,18] In contrast, intracellular maspin could affect these events directly and preliminary evidence for such a direct association will be presented later in this Chapter.

Maspin Affects Cell Growth, Angiogenesis and Apoptosis

To date, little inhibition of cell growth by mature recombinant human Maspin has been observed in vitro. This effect has however, been reported in vitro using a GST-mouse maspin fusion protein.[4] The authors demonstrate that maspin disrupts VEGF and bFGF growth factor signaling events required for the proliferation and migration of endothelial cells during angiogenesis and present compelling evidence for this mechanism of tumor growth inhibition in vivo. The details of these experiments are reported elsewhere so will not be discussed here. What is relevant here is a) that the anti-angiogenic effect appears independent of any potential protease inhibitory function that Maspin may possess; b) that maspin was present at all times during tumor growth in the animals; and c) that the result was obtained using a GST-fusion construct which likely possesses different properties to the purified mature protein when used at high concentrations. Despite these differences it would be interesting to determine what affect could be obtained on established tumors with this construct—a situation more akin to the human clinical condition. Finally, in a separate paper these same authors have demonstrated that over-expression of maspin can promote apoptosis of both pre-neoplastic and carcinomatous mammary epithelial cells in vivo—an important finding in understanding the nature of tumor suppression.[5]

Maspin is a Serpin: Implications for Function

Maspin Belongs to the "Ov-Serpin" Subgroup of Serine Proteinase Inhibitors

In 1993 Remold-O'Donnell identified a subgroup of the serpin superfamily by analyzing coding sequences of serpins that seemingly do not possess any classical secretion signals yet are found both extracellularly and intracellularly.[19] The name "ov-serpin" is derived from chicken ovalbumin (Oval) – one of the first members of this subgroup. To date there are 13 human ov-serpins which map to two independent loci on chromosomes 6p25 and 18q21 and fall into two classes based on a single difference in gene structure.[20] Maspin, plasminogen activator inhibitor 2 (PAI2), the squamous cell carcinoma antigens 1 and 2 (SCCA 1, SCCA 2), megsin, cytoplasmic antiproteinase 2 (CAP 2), and Bomapin map to 18q21.3 adjacent to the tumor promoting gene encoding *bcl* 2. The first three of these ov-serpins are involved in oncogenesis and there is mounting evidence that translocations/deletions/silencing of this locus are involved in tumor initiation and progression.[21] Recently a nomenclature meeting assigned this group to the B clade of serpins based on their "ov-serpin" subgroup classification.[22]

How do these serpins get out of the cell? Studies on oval and PAI2 have demonstrated that the uncleaved N-terminus of the protein functions as a facultative secretion signal the hydrophobicity of which determines the degree of secretion.[23] The fact that PAI-2 is found in the circulation in a glycosylated form suggests that the small amount of material that is secreted gets out via the classical secretion pathway. Our published findings have demonstrated that the N-terminus of maspin is more like that of PAI-2 suggesting that small amounts of maspin may also get out of normal cells.[16] Immunostaining studies have confirmed these findings and identified maspin at the surface of human mammary epithelial cells (HMEC's) in culture. Furthermore, other authors have demonstrated that normal cells in culture do actively secrete maspin.[24] Analysis of the primary sequence of maspin (Fig. 1) reveals 5 potential N-glycosylation sites at residues 99,133,188,260, and 361, and two potential PKC phosphorylation sites at residues 107 and 108. However, it is presently unknown what, if any, post-translational modifications the secreted form(s) of the protein possesses.

Maspin Structure-Function

Serpin Conformation and Function

Maspin is a 42 kDa protein that belongs to the serpin superfamily of Serine proteinase inhibitors. This implies that it shares conformational features that are present

```
1MDALQLANSA FAVDLFKQLC EKEPLGNVLF  SPICLSTSLS  LAQVGAKGDT50

                                                                         *
51ANEIGQVLHF  ENVKDIPFGF  QTVTSDVNKL SSFYSLKLIK  RLYVDKSLNL100

          ++                                                    *
101STEFISSTKR   PYAKELETVD FKDKLEETKG QINNSIKDLT  DGHFENILAD150

                                                                         *
151NSVNDQTKIL VVNAAYFVGK WMKKFPESET KECPFRLNKT DTKPVQMMNM200

201EATFCMGNID  SINCKIIELP   FQNKHLSMFI   LLPKDVEDES  TGLEKIEKQL250

                     *
251NSESLSQWTN  PSTMANAKVK LSIPKFKVEK MIDPKACLEN  LGLKHIFSED300

301TSDFSGMSET  KGVALSNVIH   KVCLEITEDG  GDSIEVPGAR  ILQHKDELNA350

                         *
351DHPFIYIIRH     NKTRNIIFFG     KFCSP375
```

Fig. 1. Primary sequence of maspin with potential post-translational modifications.

in all of the crystal structures of other serpins and serpin:protease complexes that have been reported to date. There are many good reviews of this subject to which the reader is referred and an abbreviated version pertinent to maspin is presented here.[20,25,26] The general serpin conformation consists of 3 β-pleated sheets (A, B, and C), and 8 α-helices. The key feature for all protease inhibitory serpins is the reactive site loop (RSL). This is a stretch of approximately 17 amino acids tethered between β-sheets A and C which exists in an exposed conformation as evidenced by its reactivity with proteases. The nature of the reactivity is largely determined by the specificity of the protease and the amino acid sequence encoded by the RSL. The critical specificity determinant is the nature of the P1 residue. With few exceptions, serpins inhibit serine proteases by an irreversible branched pathway suicide substrate inhibitory mechanism. Protease attack on the P1 residue leads to a covalent ester linkage between it and the active site serine and subsequent cleavage of the P1-P1' peptide bond. At this stage the RSL is thought to insert into β-sheet A thereby imparting enhanced stability upon the complex and disrupting protease structure to the point of rendering it inactive. The driving force for inhibition is thought to be the ability of the RSL to insert into β-sheet A—an event that also occurs following catalytic cleavage within the RSL by non-target proteases. Serpins that have no known inhibitory activity (oval, and angiotensinogen [angio] do not undergo this increase in stability and this transitional change from a stressed (S) form to a more relaxed (R) form following proteolysis has been considered a hallmark of inhibitory potential.[27-29]

Serpin	Gene Symbol	Putative RSL (P_{15}-P_5') "Proximal Hinge" … P_1	Target Protease(s)
MNEI	*SERPIN B1*	G T E A A A A T A G I A T F C M L M P E	elastase
SCCA-1	*SERPIN B3*	G A E A A A A T A V V G F G S S P A S T	cat L, S, K
SCCA-2	*SERPIN B4*	G V E A A A A T A V V V V E L S S P S T	cat G
Megsin	*SERPIN B7*	G T E A T A A T G S N I V E K Q L P Q S	
CAP-2	*SERPIN B8*	G T E A A A A T A V V R N S R C S R M E	furin
Bomapin	*SERPIN B10*	G T E A A A G S G S E I D I R I R V P S	thrombin, trypsin
Headpin	*SERPIN B13*	G T E A A A A T G I G F T V T S A P G H	
PAI-1	*SERPIN E1*	G T V A S S S T A V I V S A R M A P E E	uPA, tPA
PN1	*SERPIN E2*	G T K A S A A T T A I L I A R S S P P W	uPA
PAI-2	*SERPIN B2*	G T E A A A G T G G V M T G R T G H G G	uPA
PAI-3	*SERPIN A5*	G T R A A A A T G T I F T F R S A R L N	APC, uPA, tPA
Neuroserpin	*SERPINI1*	G S E A A A V S G M I A I S R M A V L Y	tPA
Maspin*	*SERPIN B5*	G G D S I E V P G A R I L Q H K D E L N	tPA, uPA ?
Maspin		G G D S I E V P G A - - - - R I L Q H K	

Table 1. Alignment of selected 18q21.3 Ov-serpin and other tPA/uPA inhibitory RSL sequences. RSL homology alignments. Abbreviations: MNEI-monocyte neutrophil elastase inhibitorl SCAA-squamous cell carcinoma antigen; CAP-cytoplasmic anti-proteinase; PAI-plasminogen activator inhibitor; PN-1-protease nexin 1; Cat-cathespin; uPA-urokinase type plasminogen activator; tPA-tissue type plasminogen activator; APC-activated protein C. Light grey = identical; dark grey = conservative changes; boxed = non-conservative changes.

The RSL in Maspin

In the case of maspin the RSL is significantly different from most inhibitory serpins yet is still critical for the inhibition of tumor cell invasion. Table 1 shows an alignment of "Ov-serpin" RSL sequences and RSL sequences present in other serpins known to inhibit tPA, uPA, or both proteases. It is clear that a) maspin lacks the conservation of sequence present in all other serpins within the region of the proximal hinge and, b) alignment of the RSL of maspin with the RSL sequences in other serpins is difficult as the sequence leading into the 1st strand of the C-sheet is significantly shorter than in most other serpins. Two interpretations are presented here; the first retains the continuity of the sequence through the loop without the insertion of gaps (Maspin*). The second has been published elsewhere (Maspin). The first interpretation leads to placement of Arg^{340} at the P5 position and His^{344}-Lys^{345} at the P1-P1' reactive site peptide bond. Proteolytic attack at Arg^{340} is therefore, more likely to result in catalytic cleavage of the protein than effective protease inhibition and this has been observed in vitro with trypsin-like proteases.[30] In addition, histidine as a P1 residue has never been associated with trypsin-like protease inhibitory activity—on the contrary, many "loss of function" mutations from Arg to His have been described.[12] However, the location of the reactive site peptide bond is not absolutely fixed and some serpins are able to inhibit different proteases at reactive site peptide bonds overlapping by at least 1 amino acid.[31,32] Thus, inhibition mediated by Lys^{345} is still a possibility. The second interpretation assigns Arg^{340} as the P1 residue which makes

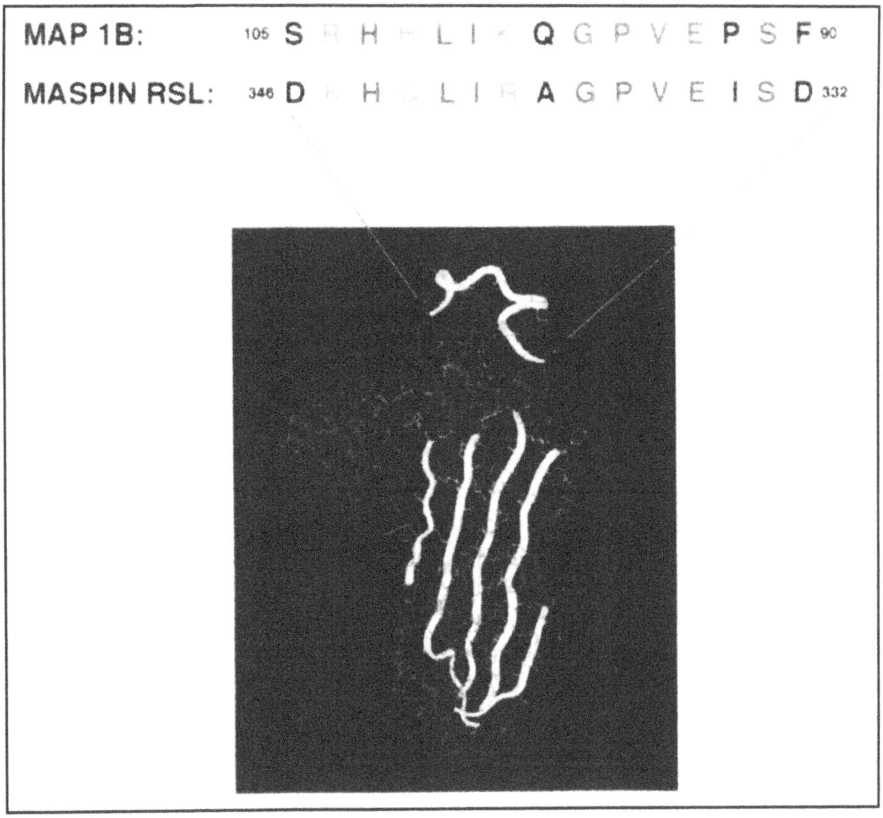

Fig. 2. Model structure of maspin: The RSL is homologous with a sequence in MAP1B.

Lys^{345} become the P5' residue. Either interpretation does not exclude the possibility of protease inhibition mediated by the RSL. Indeed, mutations within the loop at both Arg^{340} [4] and K^{345} (authors unpublished observations) or, deletions of the loop, abolish maspins ability to inhibit tumor cell migration indicating that both residues are required for this function.

Despite the inability to define a potential reactive site peptide bond by either of these approaches, the data do highlight RSL residues that are critical for function and which, incidentally, keep within the primary specificity requirements of tPA and uPA. However, purified yeast recombinant human maspin does not undergo the S to R transition in vitro and behaves more like the non-inhibitory serpins ovalbumin and angiotensinogen.[30] On this basis maspin has been modeled on the known crystal structure of ovalbumin and that structure is presented in Figure 2.[33]

Functional Insights from Homology Modeling

The maspin model reveals a distorted RSL and places the eight encoded cysteine residues at distances sufficiently distant to preclude disulphide bond formation (Table 2). Many of these also have surface topology. Titrimetric analyses with Ellman's reagent confirmed the absence of disulphide bonds in maspin and subse-

Table 2. Placement of cysteine residues in model structure of maspin

Cysteine residue number	Cysteine location
20	C-terminus of helix A
34	N-terminus of helix B
183	Middle of β-sheet strand C4
205	Between β-sheet strands C3 and B1
214	Start of β-sheet strand B2
287	Middle of helix I
323	Middle of β-sheet strand A5
373	C-terminus of β-sheet strand B5

quent studies have proven one or more of these residues to be highly reactive with alkylating reagents (authors unpublished observations). This number of free cysteine residues in a protein is unusual and suggestive of a contribution to function.

Analysis of domain subsequences encoded by maspin reveal the RSL to be almost identical to a sequence in the microtubule binding protein M AP1B (Fig. 2). However, this sequence in M AP1B is not responsible for its microtubule binding function nor is it homologous to any of those identified as required for binding in other microtubule-associated proteins (MAP's).[34,35] Despite this, fishing expeditions in mammary tumor cell lysates with immobilized recombinant human maspin or RSL-cleaved maspin as bait have identified β-tubulin as a potential intracellular target of maspin function (unpublished results). The fact that an intact RSL is not required for this function suggests that it is not directly involved in binding. What then is the significance of this homology?

Loop-Sheet Polymerization

Many intracellular serpins have been observed to undergo spontaneous polymerization in vitro but the significance of this is presently unknown. Polymerization occurs by insertion of the RSL from one molecule into the A or C sheet of another. It is induced by denaturation, mutation, or proteolysis allowing mobility of the RSL. In maspin, both native and RSL-cleaved forms of the protein can polymerize indicating access to the A-sheet is restricted following proteolysis and suggesting RSL insertion into the C-sheet to be the favored mechanism. Since β-tubulin binding does not appear to involve the RSL then maspin polymerization in vivo could function to mediate or stabilize microtubule assembly in a manner analagous to that described for M AP1B or the anti-cancer agent Taxol.[34,36] On the other hand, there is, as yet, no evidence that maspin polymerization occurs in vivo. If it doesn't, the observed binding function may serve to sequester β-tubulin in a monomeric form rendering it unavailable for the assembly of microtubules. Either potential activity would be unaffected by the integrity of the loop.

Summary of Domain Functions in Maspin

Table 3 summarizes the functions described so far and where they map on the modeled structure of maspin. The secretion of the protein appears dependent on an uncleaved facultative secretion signal encoded by the first 50 amino acids. Inhibition of tumor cell migration is dependent on the RSL whereas inhibition of endothelial cell migration and angiogenesis is not. Polymerization is dependent on the presence of

Table 3. Domain function in maspin

Domain(s)/Residues	Function(s)
N-terminal 1-50	Facultative secretion signal
Free cysteines	?
RSL	Inhibition of tumor cell invasion
	tPA/uPA inhibition
	Polymerization
?	β-tubulin binding
?	Anti-angiogenic

the RSL but not on its integrity. Binding to β-tubulin is also apparently independent of RSL integrity but as yet no binding site has been identified. The protein possesses 8 free cysteines—one or more of which are highly reactive and likely possess some functional role.

The ability to inhibit angiogenesis has been reported for several other serpins; PAI-1, pigment epithelial derived factor (PEDF), and cleaved antithrombin III (cATIII).[37-39] Inhibition by PAI-1 appears to occur by two distinct but overlapping pathways; the first is dependent on the PAI-1's protease inhibitory activity whereas the second is independent and appears to act through binding to vitronectin.[37] PEDF is responsible for the avascularity of the ocular compartments and, like maspin, its anti-angiogenic activity appears independent of protease inhibition. PEDF also has no known protease inhibitory function.[38] ATIII is the major inhibitor of thrombin activity in vivo but its anti-angiogenic activity apparently resides in a neo-epitope only present in the latent form of the protein or, in the R form generated by cleavage within the RSL. With the exception of the RSL in PAI-1, the respective sequence(s) or conformation(s) present in each of these proteins, including maspin, that confers anti-angiogenic activity has not yet been reported. Perhaps once detailed three dimensional structures are available for all of these proteins some idea as to the nature and/or similarity of these regions will become clear.

Conclusions: Future Applications—Anticipated Developments

It is clear that maspin is a multi-functional protein with identified activities and potential targets outside the cell. The identification of β-tubulin as a potential intracellular target for maspin opens up a new area of research that is particularly relevant since the bulk of maspin is cytoplasmic in normal cells. We are only just beginning to understand the nature of these interactions at a molecular level but identification of the sequences/domains responsible for these interactions, and new ones that will undoubtedly be discovered, will greatly aid our understanding of maspin's role, not only in tumor development and progression, but also in the development and maintenance of normal tissue architecture.

Finally, a multi-faceted approach for the use of maspin in the diagnosis and treatment of cancer is underway. Epidemiological efforts to clarify its role as a marker of disease progression are proceeding. Development of candidate recombinant maspins for preclinical screening in tumor bearing animal models are ongoing. The search for small molecules that reactivate the maspin gene in vivo continue. Detailed analysis of structure-function relationships within maspin will also allow development of small molecule agonists/antagonists that mimic/interfere with the individual functions of the different domains present in this novel tumor suppressor.

References

1. Zou Z, Anisowicz A, Hendrix MJC, Thor A, Neveu M, Sheng S et al. Maspin, a serpin with tumor-suppressing activity in human mammary epithelial cells. Science 1994; 263:526-529.
2. Sheng S, Pemberton PA, Sager R. Production, purifications, and characterization of recombinant maspin proteins. J Biol Chem 1994; 269:30988-30993.
3. Shao ZM, Nguyen M, Alpaugh ML, O'Connell JT, Barsky SH. The human myoepithelial cell exerts antiproliferative effects on breast carcinoma cells characterized by p21WAF1/CIP1 induction, G2/M arrest, and apoptosis. Exp Cell Res 1998; 241(2):394-403.
4. Zhang M, Volpert O, Shi Y, Bouck N. Maspin is an angiogenesis inhibitor. Nat Med 2000; 6:196-199.
5. Zhang M, Shi Y, Magit D, Furth PA, Sager R. Reduced mammary tumor progression in WAP-Tag/WAP-maspin bitransgenic mice. Oncogene 2000; 19:6053-6058.
6. Seftor RE, Seftor EA, Sheng S, Pemberton PA, Sager R, Hendrix MJ. Maspin suppresses the invasive phenotype of human breast carcinoma. Cancer Res 1998; 58:5681-5685.
7. Sheng S, Truong B, Frederickson D, Wu R, Pardee AB, Sager R. Tissue-type plasminogen activator is a target of the tumor suppressor gene maspin. Proc Natl Acad Sci USA 1998; 95:499-504.
8. McGowen R, Biliran H, Sager R, Sheng S. The surface of prsotate carcinoma DU145 cells mediates the inhibition of urokinase-type plasminogen activator by maspin.Cancer Res 2000; 60:4771-4778.
9. Sager R, Sheng S, Anisowicz A, Sotiropoulou G, Zou Z, Stenman G et al. RNA Genetics of Breast cancer: Maspin as Paradigm. Cold Spring Harb Symp Quant Biol 1994; 59:537-546.
10. Domann FE, Rice JC, Hendrix MJ, Futscher BW. Epigenetic silencing of maspin gene expression in himan breast cancers. Int J Cancer 2000; 85(6):805-810.
11. Winchester B, Vellodi A, Young E. The molecular basis of lysosomal storage diseases and their treatment. Biochem Soc Trans 2000; 28(2):150-154.
12. Carrell RW, Aulak AS, Owen MC. The molecular pathology of the serpins. Mol Biol Med 1989; 6(1):35-42.
13. Shao Z, Radziszewski WJ, Barsky SH. Tamoxifen enhances myoepithelial cell suppression of human breast carcinoma progression in vitro by two different effector mechanisms. Cancer Lett 2000; 157(2):133-144.
14. Sheng S, Carey J, Seftor E, Dias L, Hendrix MJC, Sager R. Maspin acts at the cell membrane to inhibit invasion and motility of mammary and prostatic cancerl cells. Proc Natl Acad. Sci USA 1996; 93:11669-11674.
15. Zhang M, Sheng S, Maas N, Sager R. mMaspin: the mouse homolog of a human suppressor gene inhibits mammary tumor invasion and motility. Mol Med 1997; 3(1):49-59.
16. Pemberton PA, Tipton AR, Pavloff N, Smith J, Erickson JR, Mouchabeck ZM et al. Maspin is an intracellular serpin that partitions into secretory vesicles and is present at the cell surface. J Histochem Cytochem 1997; 45:1697-1706.
17. Gomez J, Martinez A-C, Gonzalez A, Rebollo A. Dual role of Ras and Rho proteins: At the cutting edge of life and death. Immunol Cell Biol 1998; 76(2):125-134.
18. Schmitz AA, Govek EE, Bottner B, Van Aelst L. Rho GTPAses: Signaling, migration, and invasion. Exp Cell Res 2000; 262(1):1-12.
19. Remold-O'Donnell E. The ovalbumin family of serpin proteins. FEBS Lett 1993; 315(2):105-108.
20. Silverman G, Bird PI, Carrell RW, Church FC, Coughlin PB, Gettins P et al. The serpins are an expanding superfamily of structurally similar but functionally diverse

proteins: Evolution, mechanism of inhibition, novel functions, and a revised nomenclature. J Biol Chem. In press.
21. Nupponen NN, Kakkola L, Koivisto P, Visakorpi T. Genetic alterations in hormone-refractory recurrent prostate carcinomas. Am J Pathol 1998; 153(1):141-148.
22. 2nd International Symposium on the Structure and Biology of Serpins. 1999:27 June-1 July, Queens' College, Cambridge.
23. Belin D, Wohlwend A, Scheluning WD, Kruithof EKO, Vassalli JD. Facultative polypeptide translocation allows a single mRNA to encode the secreted and cystolic forms of plasminogen activators inhibitor 2. EMBO J 1989; 8:3287-3294.
24. Katz AB, Taichman LB. A partial catalog of proteins secreted by epidermal keratinocytes in culture. J Invest Dermatol 1999; 112(5):818-821.
25. Whisstock JC, Skinner R, Carrell RW, Lesk AM. Conformational changes in serpins: I. The native and cleaved conformations of alpha(1)-antitrypsin. J Mol Biol 2000; 296(2):685-699.
26. Huntington JA, Read RJ, Carrell RW. Structure of a serpin-protease complex shows inhibition by deformation. Nature 2000; 407(6806):923-926.
27. Mast AE, Enghild JJ, Pizzo SV, Salvesen G. Analysis of the plasma elimination kinetics and conformational stabilities of native, proteinase-complexed, and reactive site cleaved serpins: Comparison of alpha 1-proteinase inhibitor, alpha 1-antichymotrypsin, antithrombin III, alpha 2-antiplasmin, angiotensinogen, and ovalbumin. Biochemistry 1999; 30:2720-2728.
28. Stein PE, Tewkesbury DA, Carrell RW. Ovalbumin and angiotensinogen lack serpin S-R conformational change. Biochem J 1989; 262:103-107.
29. Gettins P, Patson PA, Schapira M. The role of conformational change in serpin structure and function. Bioessays 1993; 15(7):461-467.
30. Pemberton PA, Wong DT, Gibson HL, Kiefer MC, Fitzpatrick PA et al. The tumore suppressor maspin does not undergo the stressed to relaxed transition or inhibit trypsin-like serine proteinases. Evidence that maspin is not a protease inhibitory serpin. J Biol Chem 1995; 270(26):15832-15837.
31. Potempa J, Shieh BH, Travis J. Alpha-2-antiplasmin: A serpin with two separate but overlapping reactive sites. Science 1988; 241(4866):699-700.
32. Aulak KS, Davis AE, Donaldson VH, Harrison RA. Chymotrypsin inhibitory activity of normal C1-inhibitor and a P1 Arg to His mutant: Evidence for the presence of overlapping reactive centers. Protein Sci 1993; 2(5):727-732.
33. Fitzpatrick PA, Wong DT, Barr PJ, Pemberton PA. Functional implications of the modeled structure of maspin. Prot Engin 1996; 9:585-589.
34. Lien LL, Feener C, Fischbach N, Kunkel LM. Cloning of human microtubule-associated protein 1B and the identification of a related gene on chromosome 15. Genomics 1994; 22:273-280.
35. Ludeuna RF, Banerjee A, Khan IA. Tubulin structure and biochemistry. Curr Opin Cell Biol 1992; 4:53-57.
36. Long HJ. Paclitaxel (Taxol): A novel anticancer chemotherapeutic drug. Mayo Clin Proc 1994; 69(4):341-345.
37. Steffansson S, Petitclerc E, Wong MK, McMahon GA, Brooks PC, Lawrence DA. Inhibition of angiogenesis in vivo by plasminogen activator inhibitor-1. J Biol Chem 2000; 32(5):8135-8141.
38. Dawson DW, Volpert OV, Gillis P, Crawford SE, Xu H, Benedict W et al. Pigment epithelium-derived factor: A potent inhibitor of angiogenesis. Science 1999; 285(5425):245-248.
39. O'Reilly MS, Pirie-Shepherd S, Lane WS, Folkman J. Antoangiogenic activity of the cleaved conformation of the serpin antithrombin. Science 1999; 285(5435):1926-1928.
40. Impagnatiello MA, Bianchi E, Di Stefano R, Pardi R, Mosca F. A novel approach to the identification of genes involved in neo-angiogenesis: Implications for graft re-vascularization. Transplant Proc 1997; 29(1-2):1110.
41. Xia W, Lau Y-K, Hu MC-T, Li L, Johnston DA, Sheng S et al. High tumoral maspin expression is associated with improved survival of patients with oral squamous cell carcinoma. Oncogene 2000; 19:2398-2403.

Chapter 4

Maspin and Myoepithelial Cells

Sanford H. Barsky, Paul Kedeshian and Mary L. Alpaugh

Introduction

Host cellular paracrine regulation of tumor progression is an important determinant of tumor growth, invasion and metastasis but one cell which has largely been ignored in this regulation is the myoepithelial cell. In any organ where there is significant branching morphogenesis such as the breast, myoepithelial cells ubiquitously accompany and surround epithelial cells and are thought to keep in check (negatively regulate) the process of branching. Myoepithelial cells surround both normal ducts and precancerous lesions, especially of the breast (so-called DCIS, ductal carcinoma-in-situ), and form a natural border separating proliferating epithelial cells from proliferating endothelial cells (angiogenesis). Myoepithelial cells, by forming this natural border, are thought to negatively regulate tumor invasion and metastasis. Whereas epithelial cells are susceptible targets for transforming events leading to cancer, myoepithelial cells are resistant. Indeed tumors of myoepithelial cells are uncommon and when they do occur, are almost always benign. Therefore it can be said that myoepithelial cells function as both autocrine as well as paracrine tumor suppressors. Our laboratory has found that myoepithelial cells secrete a number of suppressor molecules including high amounts of diverse proteinase inhibitors which include TIMP-1, protease nexin-II, and α-1 antitrypsin, but low amounts of proteinases and high amounts of diverse angiogenic inhibitors which include thrombospondin-1 and soluble bFGF receptors but low amounts of angiogenic factors compared to common malignant cell lines. However the most striking difference between the suppressive effector molecules secreted by myoepithelial cells and carcinoma cells is the levels of maspin secretion. Whereas carcinoma cells do not secrete maspin, myoepithelial cells secrete this serpin in large quantities. This observation holds in vitro, in mice, and in humans and suggests that maspin and myoepithelial cells exert pleiotropic suppressive effects on tumor progression. Since maspin is both a proteinase inhibitor, a locomotion inhibitor and an angiogenesis inhibitor, the diverse actions of maspin may largely explain the pronounced anti-invasive and anti-angiogenic effects of myoepithelial cells on carcinoma and pre-carcinoma cells. The same actions of maspin also may account for the low grade biology of myoepithelial tumors which are devoid of appreciable angiogenesis and invasive behavior. Finally since maspin is a secretory product of myoepithelial cells, the presence of maspin in body fluids such as in breast ductal fluid and in saliva reflects the structural and functional integrity of the

Maspin, edited by Mary J.C. Hendrix. ©2002 Eurekah.com

ductal-lobular units of the mammary and salivary glands respectively. Maspin, in ductal fluid, may serve as a surrogate (intermediate) end point marker (SEM) to estimate the risk of DCIS progression to invasive cancer in the breast and alternatively, in saliva, may serve as a tumor marker to detect the presence of incipient myoepithelial tumors occurring within the salivary glands of the head and neck.

Review of Current Research

Studies of Maspin and Myoepithelial Cells

It has become clear that cancer cells come under the influence of important paracrine regulation from the host microenvironment.[1] Such host regulation may be as great a determinant of tumor cell behavior in vivo as the specific oncogenic or tumor suppressor alterations occurring within the malignant cells themselves, and may be mediated by specific extracellular matrix molecules, matrix-associated growth factors, or host cells themselves.[2,3] Both positive (fibroblast, myofibroblast and endothelial cell) and negative (tumor infiltrating lymphocyte and cytotoxic macrophage) cellular regulators exist which profoundly affect tumor cell behavior in vivo.[4,5] One host cell however, the myoepithelial cell, has escaped the paracrine onlooker's attention. The myoepithelial cell, which lies on the epithelial side of the basement membrane, is thought to contribute largely to both the synthesis and remodeling of this structure. This cell lies in juxtaposition to normally proliferating and differentiating epithelial cells in health and to abnormally proliferating and differentiating epithelial cells in precancerous disease states such as ductal carcinoma in situ (DCIS) of the breast. This anatomical relationship suggests that myoepithelial cells may exert important paracrine effects on normal glandular epithelium and may regulate the progression of DCIS to invasive breast cancer. Circumstantial evidence suggests that the myoepithelial cell naturally exhibits a tumor suppressor phenotype. Myoepithelial cells rarely transform and when they do generally give rise to benign neoplasms that accumulate rather than degrade extracellular matrix.[6] Myoepithelial cells directly or indirectly through their production of extracellular matrix and proteinase inhibitors including maspin are thought to regulate branching morphogenesis that occurs in the developing breast and salivary gland during embryological development.[7] There have been a paucity of studies on myoepithelial cells because they have been relatively difficult to culture and because tumors that arise from these cells and rare.

In previous studies we have been extremely fortunate to have successfully established immortalized cell lines and transplantable xenografts from benign or low grade human myoepitheliomas of the salivary gland and breast.[8,9] These cell lines and xenografts displayed an essentially normal diploid karyotype and expressed identical myoepithelial markers as their in situ counterparts including high constitutive expression of maspin. Unlike the vast majority of human tumor cell lines and xenografts which exhibited matrix-degrading properties, these myoepithelial lines/xenografts like their myoepithelial counterparts in situ retained the ability to secrete and accumulate an abundant extracellular matrix composed of both basement membrane and non-basement membrane components. When grown as a monolayer one prototype myoepithelial cell line, HMS-1, exerted profound and specific effects on normal epithelial and primary carcinoma morphogenesis.[8] These studies support our position that our established myoepithelial lines/xenografts recapitulate a normal differentiated myoepithelial phenotype and can therefore be used experimentally as a primary myoepithelial surrogate. Prompted by these studies and by the conspicuous absence of studies examining the role of the myoepithelial cell in tumor progression, we decided to examine the myoepithelial cell from this perspective. Experiments with these cell lines/xenografts together with relevant in situ observations form the cornerstone of our studies which observe that the human myoepithelial cell is a natural tumor suppressor.

Maspin and Myoepithelial Cells Inhibit Tumor Invasion

Breast ducts and acini are surrounded by a circumferential layer of myoepithelial cells exhibiting strong immunoreactivity for S100, smooth muscle actin, and diverse proteinase inhibitors including maspin, α1-AT, PNII/APP, and TIMP-1 (Fig.1A). In DCIS, the myoepithelial layer appeared either intact or focally disrupted, but the myoepithelial cells themselves exhibited the same pattern of immunoreactivity (Fig. 1B). In DCIS although proliferations of vWf immunoreactive blood vessel capillaries were observed focally within the supporting stroma, such blood vessels were not observed within the proliferating DCIS cells on the epithelial side of the myoepithelial layer (Fig.1C). The human tumoral-nude mouse xenografts derived from the human myoepitheliomas of the salivary gland, HMS-X and HMS-3X, and breast, HMS-4X demonstrated an immunocytochemical profile identical to each other and to that exhibited by the myoepithelial cells surrounding normal ducts and DCIS with especially intense maspin immunoreactivity (Fig.1D). Not only was strong proteinase inhibitor immunoreactivity present within the myoepithelial cells of these xenografts, but strong proteinase inhibitor immunoreactity could be demonstrated within their extracellular matrix as well. Within this abundant extracellular matrix deposited by the different human myoepithelial xenografts, murine blood vessels were not observed (Fig. 1E, Fig.1F; Fig 1G). Using a mouse specific Cot-1 DNA probe (Fig. 1H), human myoepithelial xenografts HMS-X, HMS-3X, and HMS-4X, demonstrated an absent or near absent murine component indicative of absent to near absent angiogenesis. Human non-myoepithelial xenografts of breast cancer cell lines, MDA-MB-231 and MDA-MB-468, in contrast, showed a comparatively large murine component of presumed angiogenesis. The human myoepithelial xenografts, HMS-X, HMS-3X, and HMS-4X, instead exhibited a very dense accumulation of extracellular matrix accounting for a gross "pearl" appearance of these xenografts (Fig. 1I)

Detailed studies[9] conducted with HMS-1 (Fig. 1J), a prototype myoepithelial cell line, revealed a constitutively high proteinase inhibitor to proteinase ratio in strong contrast to the high proteinase to proteinase inhibitor ratio observed in a number of malignant human cell lines (Fig. 2A). Marker studies with this cell line and corresponding xenograft (HMS-X) reflected the constitutive gene expression profile of myoepithelial cells in situ (Fig. 2B). This was especially true with respect to maspin. Direct gelatin zymography of CM revealed only low levels of the 92 and 72 kDa type IV collagenases (MMP-9 and MMP-2 respectively) in HMS-1; the 72 kDa collagenase was reduced 6-fold in HMS-1 compared to the levels in the majority of the malignant lines; direct fibrin zymography revealed visibly lower levels of the 54 kDa urokinase plasminogen activator (uPA) in HMS-1. This was also observed in casein/plasminogen gels. Tissue type plasminogen activator was not detected in any cell line, nor was plasmin detected in control gels lacking plasminogen. Stromelysin-1 (MMP-3) was

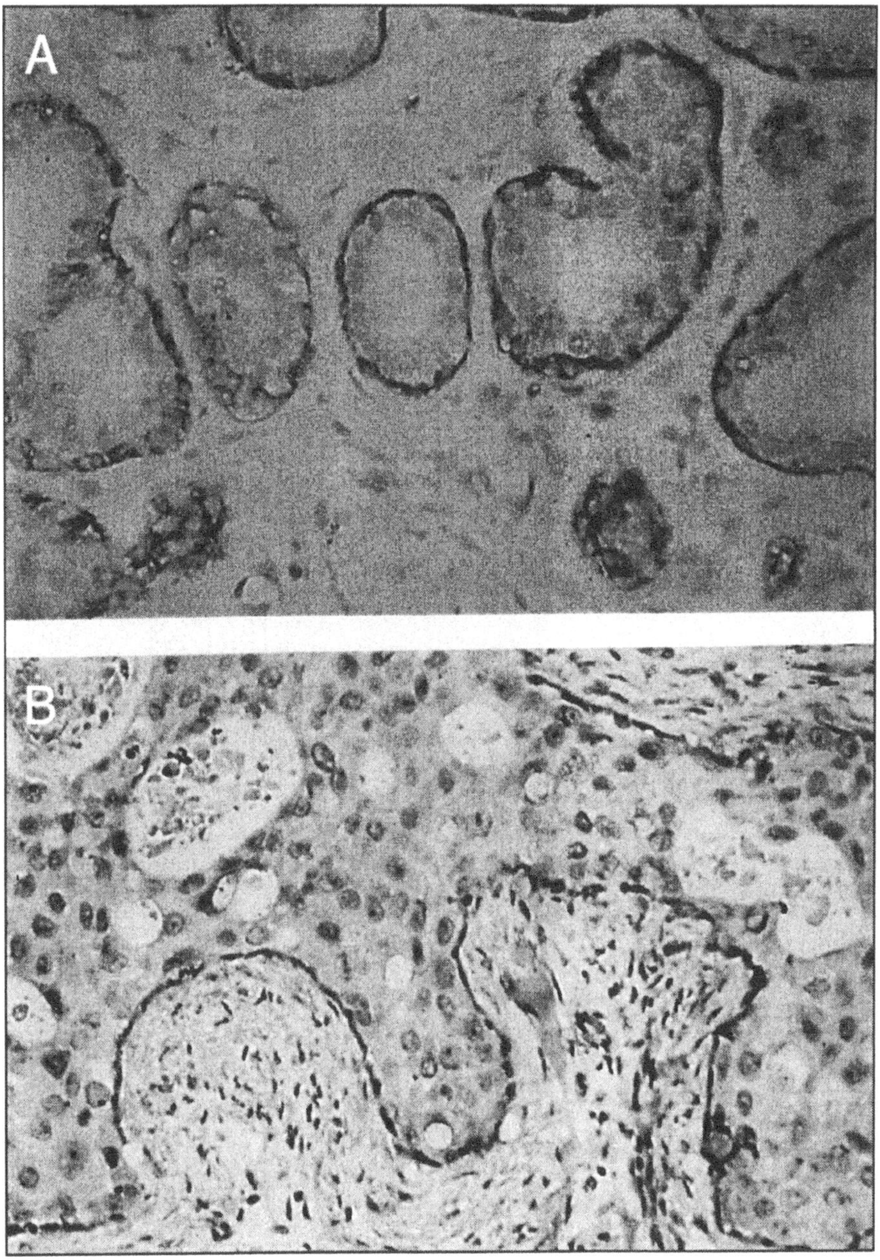

Fig. 1A & B. In situ immunocytochemistry profile of myoepithelial cells and their derived cell lines/ xenografts. (A) Differential maspin immunoreactivity of myoepithelial cells surrounding breast ducts and acini; (B) Differential maspin immunoreactivity of myoepithelial cells in DCIS.

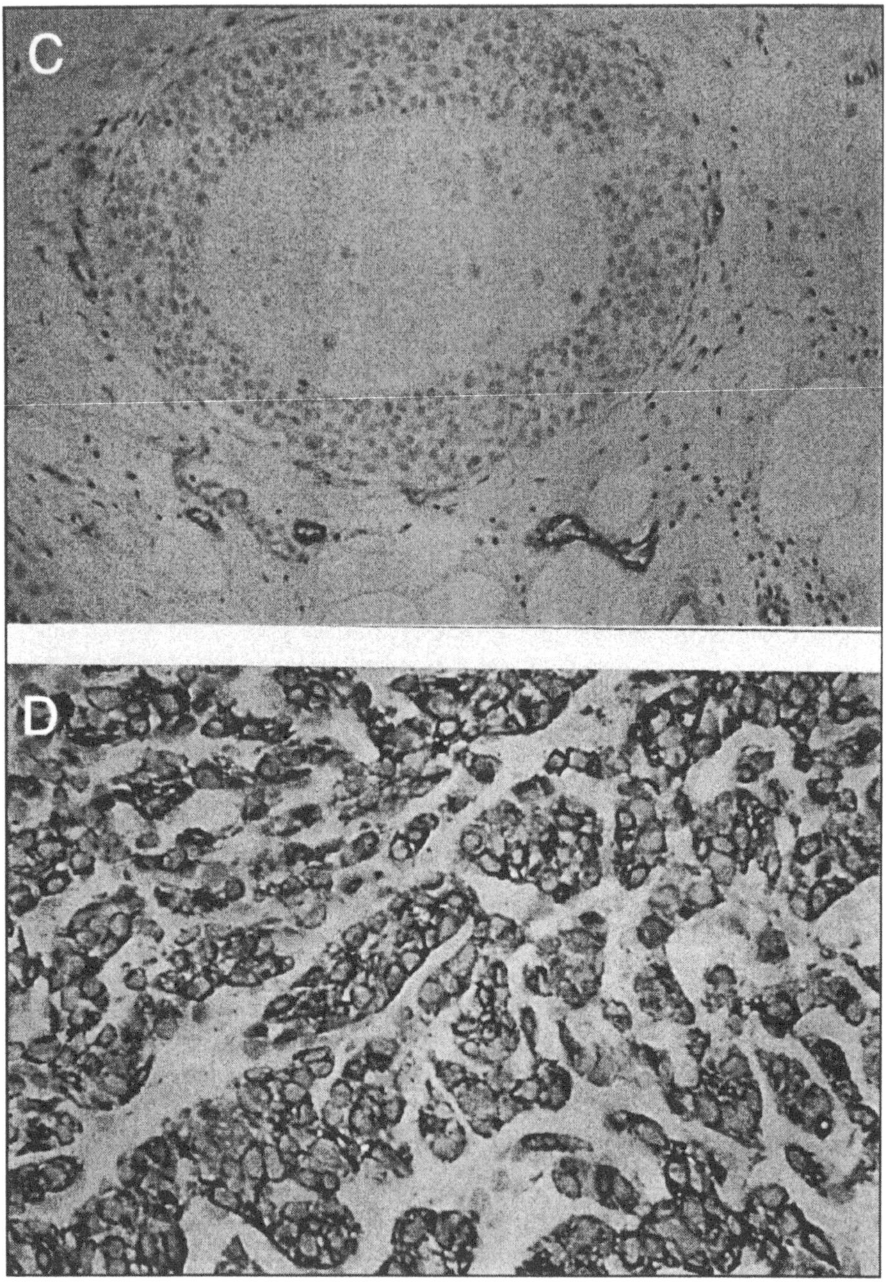

Fig. 1C & D. In situ immunocytochemistry profile of myoepithelial cells and their derived cell lines/xenografts. (C) Angiogenesis demonstrated by vWf immunoreactivity limited to stromal side of DCIS; (D) Cytoplasmic maspin immunoreactivity of myoepithelial xenograft, HMS-X.

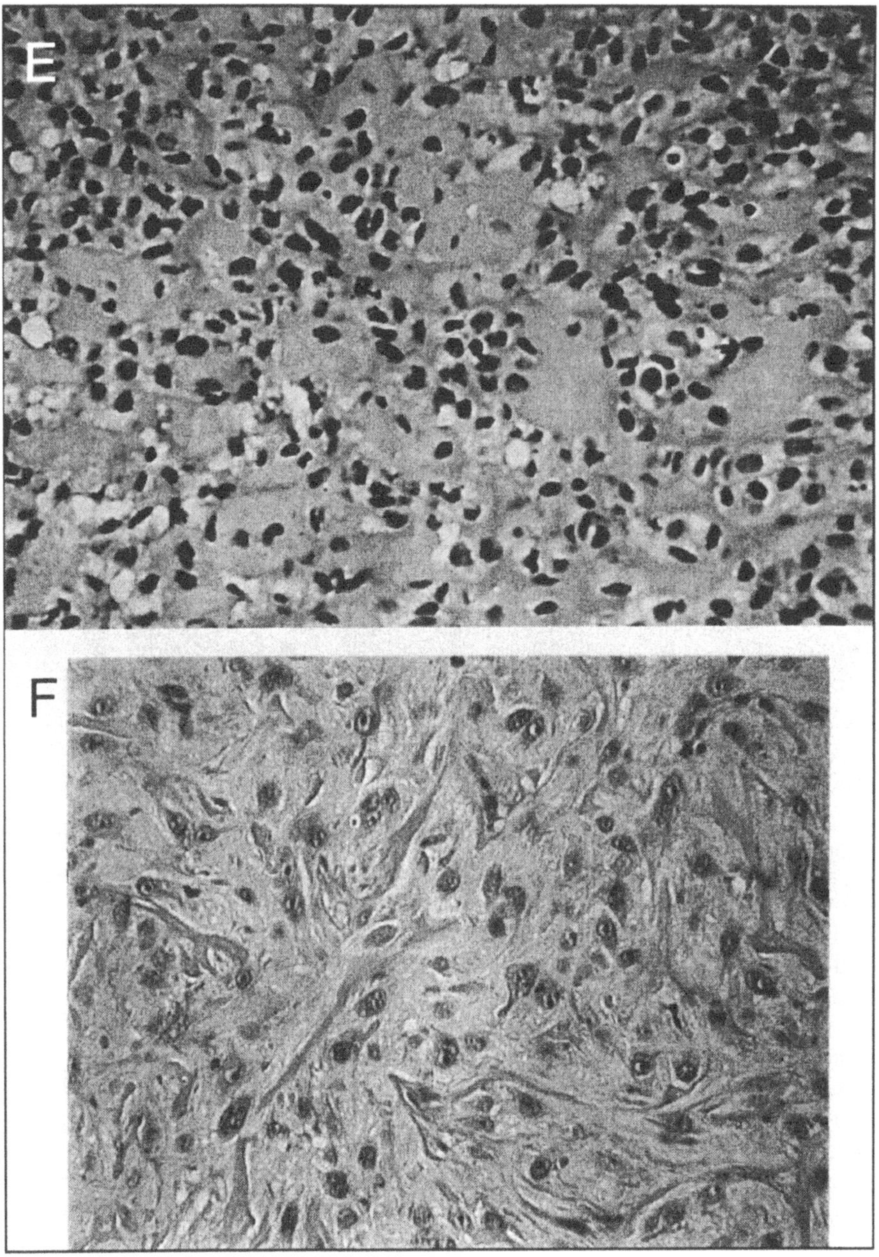

Fig. 1E & F. In situ immunocytochemistry profile of myoepithelial cells and their derived cell lines/ xenografts. (E) HMS-X, (F) HMS-4X.

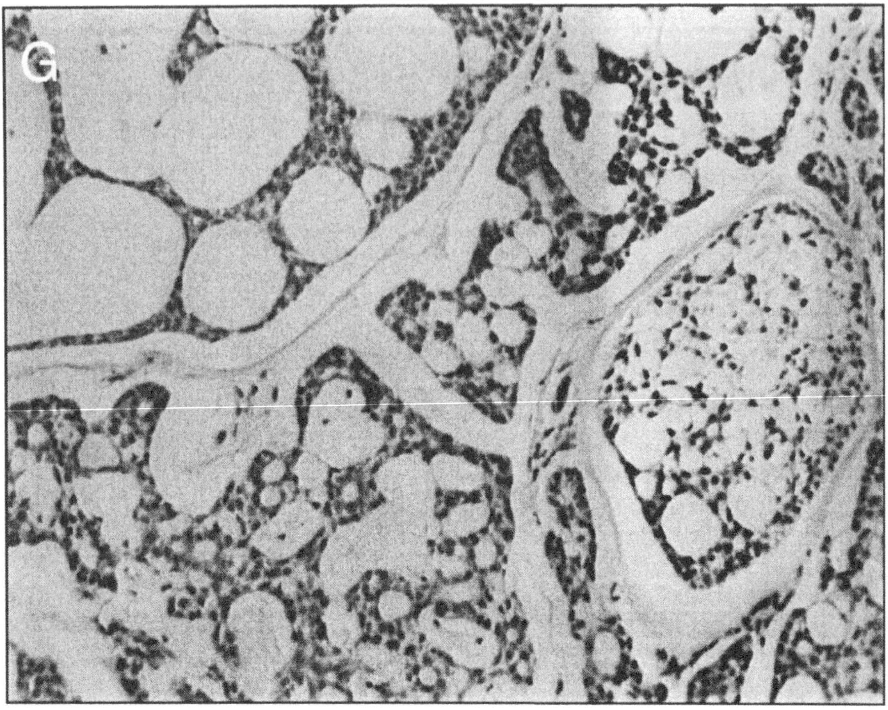

Fig. 1G. In situ immunocytochemistry profile of myoepithelial cells and their derived cell lines/ xenografts. HMS-6X myoepithelial xenografts exhibit abundant matrix accumulation devoid of apparent angiogenesis.

also not detected in HMS-1. The proteinase inhibitor expression profile of HMS-1, in contrast, was characterized by high constitutive expression in CM of several proteinase inhibitors including TIMP-1; PAI-1; three trypsin inhibitors: α1-AT, PNII/APP, and an unidentified 31 kDa inhibitor detected initially on reverse zymography; and the tumor suppressor maspin. With respect to the trypsin serine proteinase inhibitors, the conspicuous doublet at 116 kDa consistently greater in HMS-1 than in any of the other lines examined was confirmed on Western blot to be PNII/APP. These bands represented the 770 and 751 amino acid isoforms of PNII/APP which possessed a Kunitz-type serine proteinase inhibitor domain. Interestingly in 2M urea extracts of HMS-X, HMS-3X, and HMS-4X, a novel 95 kDa band of trypsin inhibition was detected by reverse zymography and confirmed by Western blot to represent an active breakdown product of PNII. This 95 kDa PNII breakdown product was completely absent from HMS-1 CM and urea extracts of HMS-1 cells suggesting that it was produced in situ within the myoepithelial extracellular matrix to which it bound. The retention of proteinase inhibitor activity by this breakdown product indicated that it retained the Kunitz-type serine proteinase inhibitor domain responsible for its ability to inhibit trypsin. In contrast to PNII/APP, protease nexin I was not detected. The second trypsin serine proteinase inhibitor was present at 54 kDa and was α1-AT. This inhibitor appeared nearly equivalent in HMS-1 compared to the malignant lines examined on reverse zymography, but by Western blot its signal was markedly stronger and slightly more mobile in HMS-1 than in the malignant lines. This data was

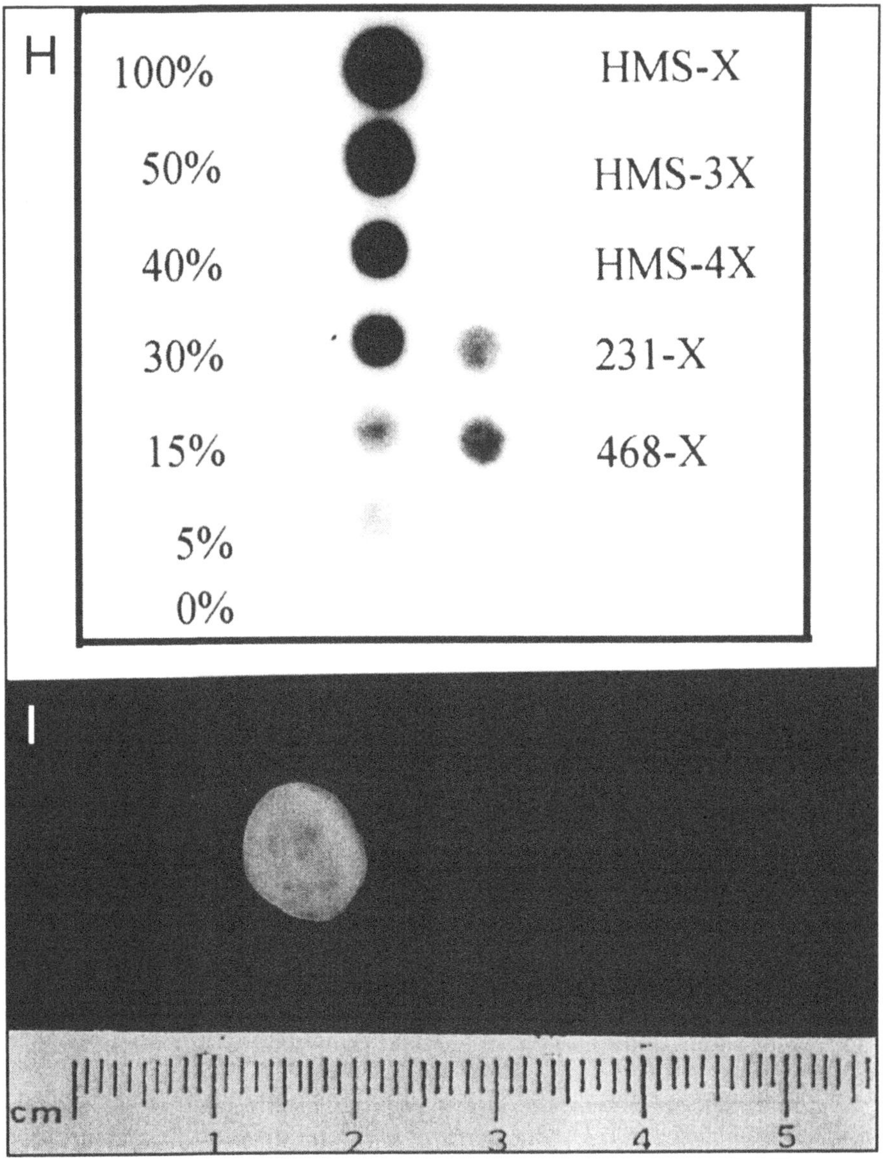

Fig. 1H & I. In situ immunocytochemistry profile of myoepithelial cells and their derived cell lines/xenografts. (H) Murine Cot-1 dot blot. Using a mouse specific Cot-1 DNA probe, human myoepithelial xenografts, HMS-X, HMS-3X and HMS-4X, are devoid of a murine DNA angiogenic component in contrast to the angiogenic-rich MDA-MB-231-X and MDA-MB-468-X breast carcinoma xenografts (right column); control dot blots of varying murine DNA percentages are also depicted (left column); (I) The typical myoepithelial xenograft looks and shells out like a white glistening "pearl".

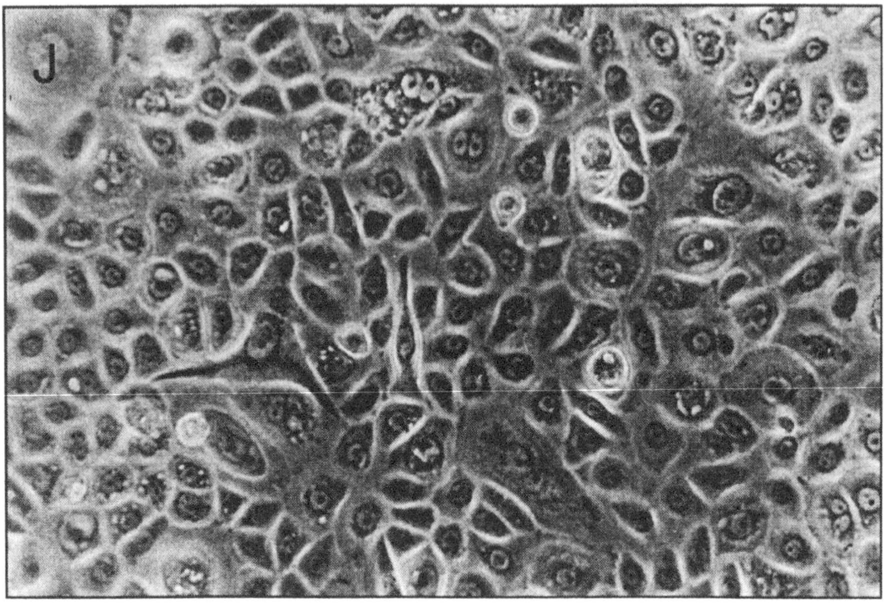

Fig. 1J. In situ immunocytochemistry profile of myoepithelial cells and their derived cell lines/ xenografts. Myoepithelial cells (HMS-1) grow as a monolayer in cell culture.

reconciled with the fact that α1-AT was probably less glycosylated in HMS-1. This relative underglycosylation caused α1-AT from HMS-1 to migrate slightly further into the gel and accounted for its poorer reactivation following SDS-denaturation on reverse zymography as compared to the more highly glycosylated isoforms present in the malignant lines. The third trypsin serine proteinase inhibitor detected at 31 kDa was clearly not a degradation product of either PNII or α1-AT as demonstrated by negative Western blot. The 31 kDa inhibitor was strongly expressed in HMS-1 and was either absent or nearly absent in all of the malignant lines examined. Whether this unidentified inhibitor is a novel inhibitor is being determined. In contrast to the above inhibitors, PAI-1 was expressed only slightly greater in HMS-1 compared to the majority of the malignant lines by both reverse zymographic and Western blot analysis. Neither PAI-2, PAI-3 or α2-antiplasmin were detected by Western blot analysis in any of the cell lines. Antiplasmin activity as determined by photometric assay was completely absent as well. The most striking difference, however, between the strong proteinase inhibitor profile of HMS-1 and the profile of the malignant cell lines examined was in the expression of maspin. Intense maspin transcripts (3.0 and 1.6 kb) and protein (42 kDa) were identified in HMS-1 and HMS-1 CM respectively but were completely absent in all of the malignant lines examined (Fig. 2A). With its proteinase inhibitor profile of increased maspin, TIMP-1, PNII, α1-AT and the 31 kDa inhibitor, HMS-1 bore strong resemblance to normal human mammary epithelial cells (HMEC) (Fig. 2A) except that the expression of all of these proteinase inhibitors including maspin was even more enhanced in HMS-1. Being derived from normal ducts and acini of the human breast, HMEC cultures likely contain myoepithelial as well as epithelial cells. Thus the resemblance of HMS-1 to HMEC further supported our contention that HMS-1, though immortal, expressed a well-differentiated myo-

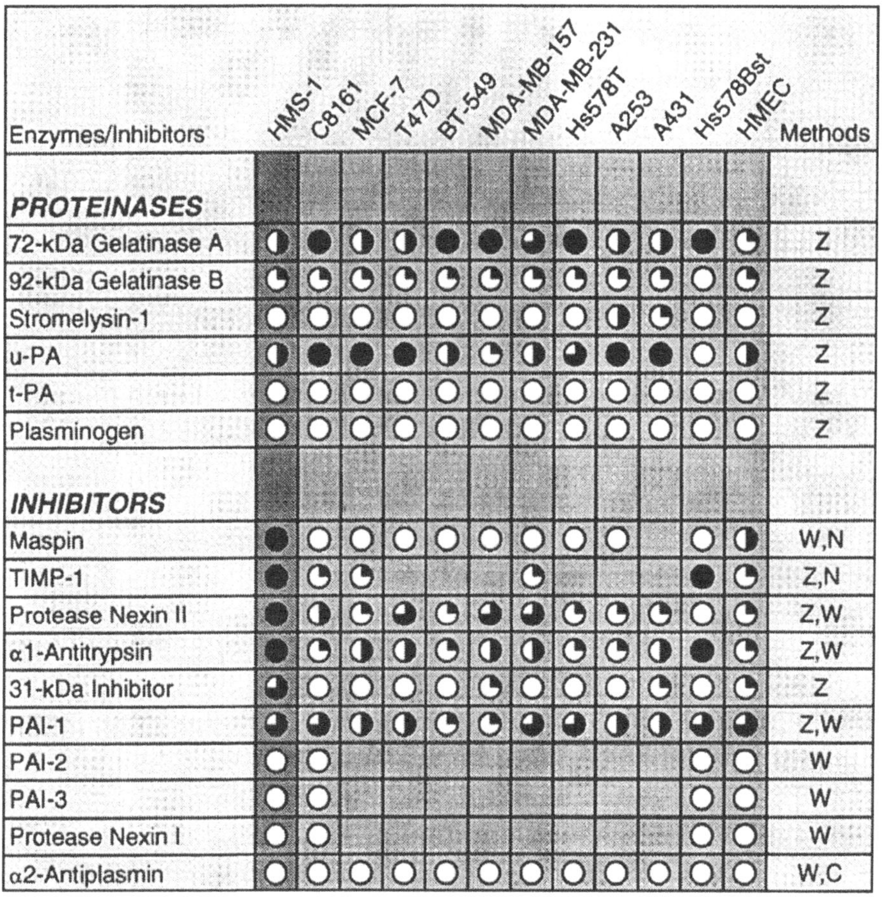

Fig. 2A. Relative constitutive expression of diverse proteinase inhibitors and proteinases in myoepithelial cells (HMS-1) compared to various malignant cell lines. Z, direct or reverse zymography; W, western blot; N, northern blot; C, chromogenic substrate assay.

epithelial phenotype. In addition, since HMS-1 was a clonal line expressing a pure myoepithelial phenotype, it would be predicted to express certain myoepithelial-associated proteins such as maspin, α1-AT, PNII/APP, and TIMP-1 to a greater degree than HMEC. Predictably, the myofibroblast line, Hs578Bst, was strongly expressive of TIMP-1 but did not express maspin, PNII or the 31 kDa inhibitor (Fig 2A). The strong proteinase inhibitor profile exhibited by HMS-1 was shared by all of the myoepithelial xenografts including HMS-X, HMS-3X, and HMS-4X.

In the modified Matrigel invasion chamber used in this study, HMS-1 cells and their conditioned media (CM) dramatically inhibited invasion of four invasive breast carcinoma cell lines (Fig. 2C). The HMS-1 line was itself non-invasive in this chamber. Predictably, the anti-invasive effects of HMS-1 could be abolished by CHX (40 μg/ml) 24 hr pretreatment. HMS-1 CM inhibited invasion in a dose response fashion up to 30%±8% of control ($p<.01$). Pretreatment of HMS-1 with dexamethasone (.25 μM) produced a complete invasion-permissive phenotype (100% of control) whereas

	HMS-X, 3X, 4X		Normal Breast		DCIS	
	Cells	Matrix	ME*	Epi‡	ME	Epi
S-100	++++§	-	++++	-	++++	-
Maspin	++++	-	++++	+	++++	±
α1-AT	++	++	++	-	++	-
PNII	++	+++	++	-	++	±
TIMP-1	++	+	+	-	++	-
PAI-1	+	±	+	+	+	+

Fig. 2B. Myoepithelial-related immunoreactivity in situ.
* myoepithelial cells; ‡ epithelial cells; § ++++, intensely positive; +++, strongly positive;++, positive; +, weakly positive; ±, equivocally positive; -, negative.

pretreatment with PMA (5 μM) produced an essentially nonpermissive phenotype (2% of control) (p<.05) (Fig. 2D). The effects of dexamethasone and PMA were quite dramatic. The effects of other agents including RA, dB-cAMP, Na-But, and 5-AzaC showed either permissive or non-permissive trends but were less dramatic. PMA's induction of the nonpermissive phenotype began after 20 minutes pretreatment, was almost complete after 2 hr and maximized after 24 hr (p<.05). The induction of this nonpermissive phenotype correlated with the induction of a dramatic 5-fold increase in maspin secretion measured in HMS-1 CM (Fig. 2E). As a result of PMA treatment, both an immediate release (within 2 minutes) of maspin from HMS-1 cells occurred (Fig. 2E) as well as a more sustained secretion for at least 24 hr following PMA pretreatment (Fig. 2E). The increased maspin secretion was not on the basis of an increase in steady state maspin transcripts (Fig. 2E). PMA also resulted in a less dramatic 2-fold increase in both MMP-9 and TIMP-1 secretion. Dexamethasone's induction of an invasion-permissive phenotype in HMS-1 was not associated with a change in either maspin transcription or secretion. Immunoprecipitation with anti-maspin antibody at 1/100 dilution successfully removed all detectable maspin from myoepithelial cell CM (Fig. 2F). This CM lost its ability to inhibit invasion (Fig. 2F). Similar results were observed with the CM from the other myoepithelial lines (HMS-3 and HMS-4) studied. None of the non-myoepithelial cell CM inhibited invasion.

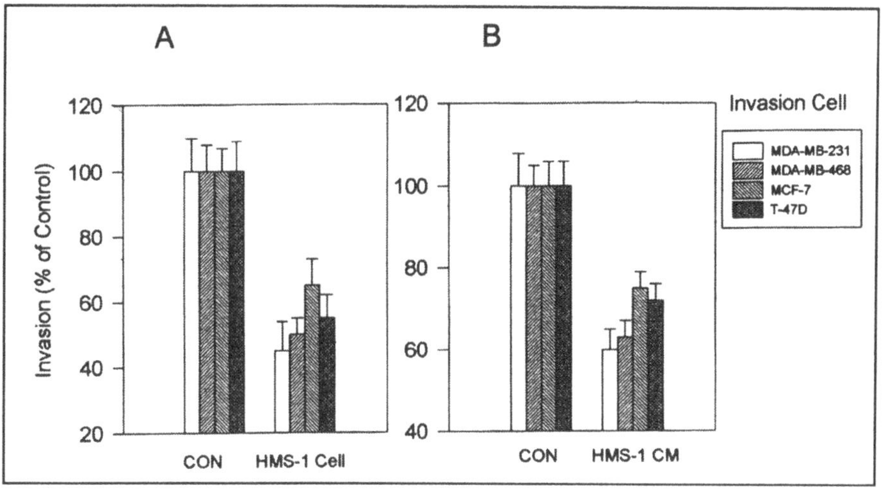

Fig. 2C. Effects of HMS-1 cells [A] and conditioned media (CM) [B] on MCF-7, T47D, MDA-MB-468 and MDA-MB-231 invasion. Both assays were performed in quadruplicate and show mean % invasion ± standard deviation.

Maspin and Myoepithelial Cells Inhibit Tumor Angiogenesis

Human myoepithelial cells which surround ducts and acini of certain organs such as the breast form a natural border separating epithelial cells from stromal angiogenesis. Myoepithelial cell lines (HMS-1-6), derived from diverse benign myoepithelial tumors, all constitutively express high levels of active angiogenic inhibitors which include maspin, TIMP-1, thrombospondin-1 and soluble bFGF receptors but very low levels of angiogenic factors.[10] Maspin recently has been shown conclusively to be an angiogenesis inhibitor.[11] As expected, our myoepithelial cell lines inhibit endothelial cell chemotaxis and proliferation. These myoepithelial cell lines sense hypoxia, respond to low O_2 tension by increased HIF-1α but with only a minimal increase in VEGF and iNOS steady state mRNA levels. Their corresponding xenografts (HMS-X-6X) grow very slowly compared to their non-myoepithelial carcinomatous counterparts and accumulate an abundant extracellular matrix devoid of angiogenesis but containing bound angiogenic inhibitors. These myoepithelial xenografts exhibit only minimal hypoxia but extensive necrosis in comparison to their non-myoepithelial xenograft counterparts. These former xenografts inhibit local and systemic tumor-induced angiogenesis and metastasis presumably from their matrix-bound and released circulating angiogenic inhibitors. These observations collectively support the hypothesis that the human myoepithelial cell (even when transformed) is a natural suppressor of angiogenesis.

Myoepithelial cells in situ separate epithelial cells from stromal angiogenesis, and this seemingly banal observation serves to illustrate the fact that stromal angiogenesis never penetrates this myoepithelial barrier (Fig. 1C) raising the hypothesis that myoepithelial cells are natural suppressors of angiogenesis. This observation was reinforced by a microscopic, immunohistochemical and DNA analysis of our myoepithelial xenografts. Our diverse myoepithelial xenografts secrete and accumulate an abundant extracellular matrix which is devoid of blood vessels in routine hematoxylin

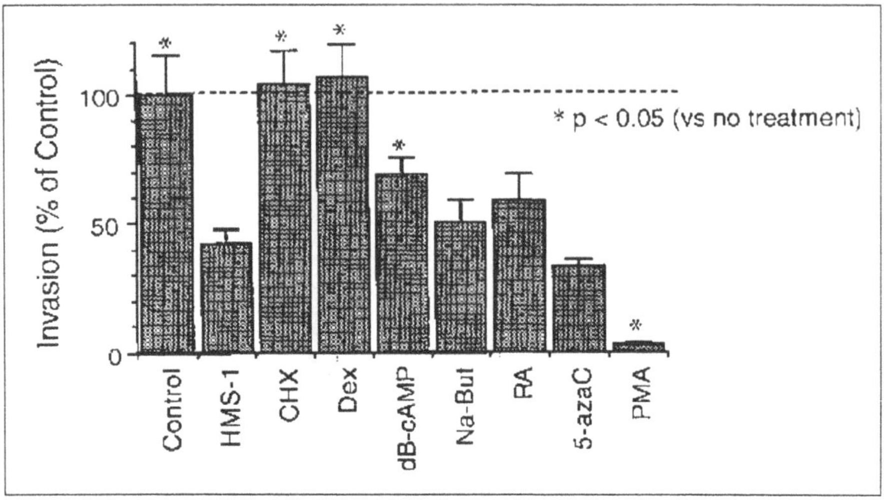

Fig. 2D. Effects of pharmacologic treatment of HMS-1 cells with various agents inducing permissive and non-permissive phenotypes: CHX, cyclohexamide; DEX, dexamethasone; dB-cAMP, N^6,2'-O-dibutyryladenosine 3':5'-cyclic monophosphate; Na-But, sodium butyrate; RA, all *trans* retinoic acid; 5-azaC, 5-azacytidine; PMA, phorbol 12-myristate 13-acetate.

and eosin staining (Fig. 1E; Fig. 1F; Fig. 1G) and vWf immunocytochemical staining in contrast to non-myoepithelial xenografts which show bursts of blood vessels. Quantitation of vessel density in 10 HPFs reveals absent to low vessel density in the myoepithelial xenografts compared to the non-myoepithelial xenografts ($p<0.01$). As mentioned previously, murine DNA Cot-1 analysis further reveals the absence of a murine component in the myoepithelial xenografts. Since in the xenografts, angiogenesis would be murine in origin, the absence of a murine DNA component is another indication that angiogenesis is minimal. Interestingly the myoepithelial xenografts grew slowly compared to the non-myoepithelial xenografts, a feature which was not found in comparisons between the myoepithelial *v* non-myoepithelial cell lines.

To explain these in vivo observations, we analyzed the gene expression profiles of our myoepithelial cell lines *v* non-myoepithelial cell lines with respect to known angiogenic inhibitors and angiogenic factors. HMS-1, as a prototype myoepithelial cell line, constitutively expressed none of the known angiogenic factors including bFGF, aFGF, angiogenin, TFGα, TGFβ, TNF-α, VEGF, PD-ECGF, PIGF, IFα, HGF, and HB-EGF but rather expressed maspin, thrombospondin-1, TIMP-1 and soluble bFGF receptors at high levels; this was in contrast to a high angiogenic factor (which included bFGF, VEGF, TFGα, TGFβ, HB-EGF, and PD-ECGF) to angiogenic inhibitor gene expression profile which was observed in non-myoepithelial cell lines. Other myoepithelial cell lines (HMS-2-6) exhibited an angiogenic inhibitor/angiogenic factor profile similar to that of HMS-1. Interestingly in 2M urea extracts of the myoepithelial xenografts but not in any of the non-myoepithelial xenografts, strong thrombospondin-1, TIMP-1 as well as plasminogen and prolactin fragments could be detected by Western blot. HMS-1 and HMS-1 CM (concentrated 10-100 fold) exerted a marked inhibition of endothelial migration and proliferation, both of which were abolished by pretreatment of the myoepithelial cells with cyclohexamide or dexamethasone. HMS-1 cells themselves did not migrate in response to either K-SFM,

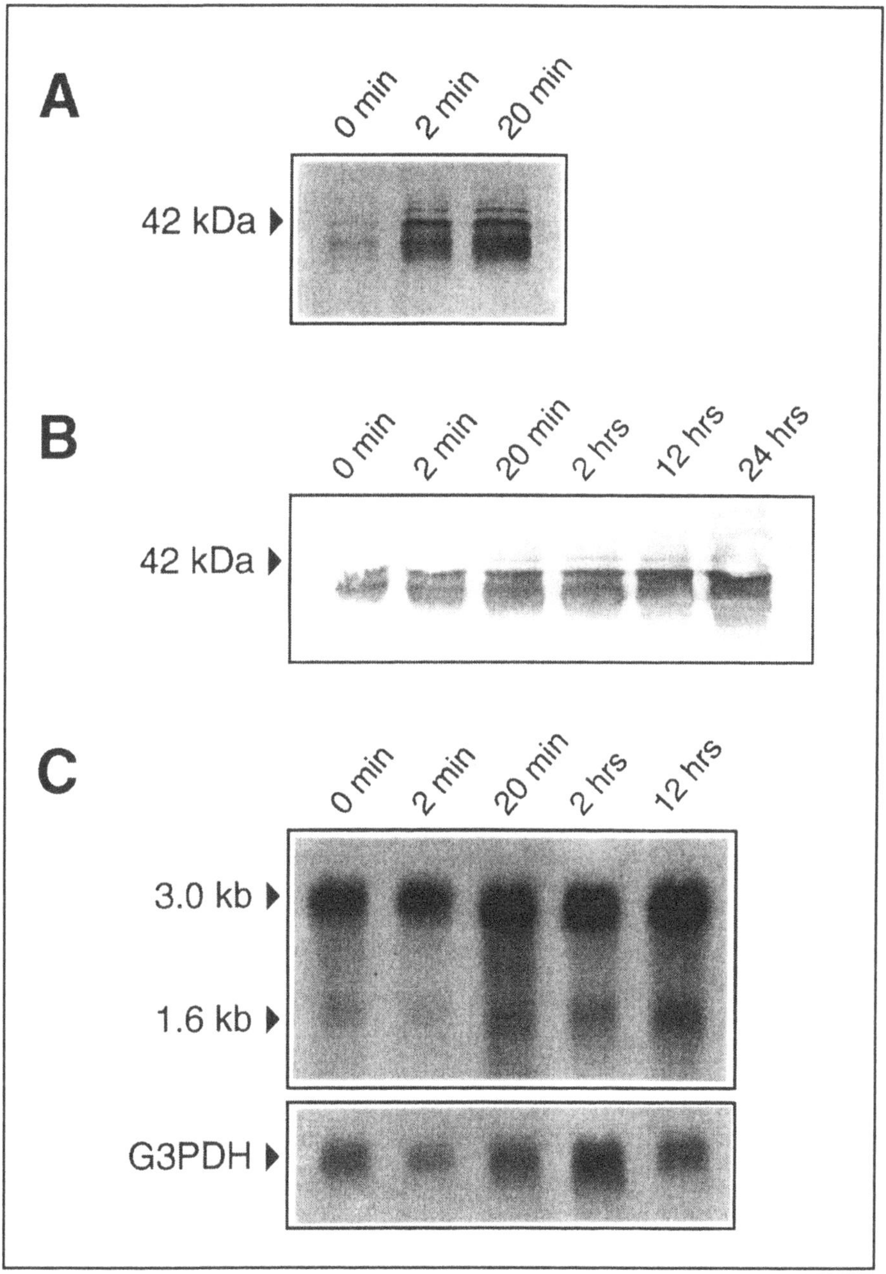

Fig. 2E. Immediate effects of PMA treatment on maspin secretion measured by Western blot of CM at designated times of PMA exposure [A]; delayed effects of PMA on maspin secretion measured by Western blot of CM 72 hours after PMA pretreatment for indicated time periods followed by removal of PMA [B]; Northern blot of maspin expression following exposure to PMA for indicated time periods [C].

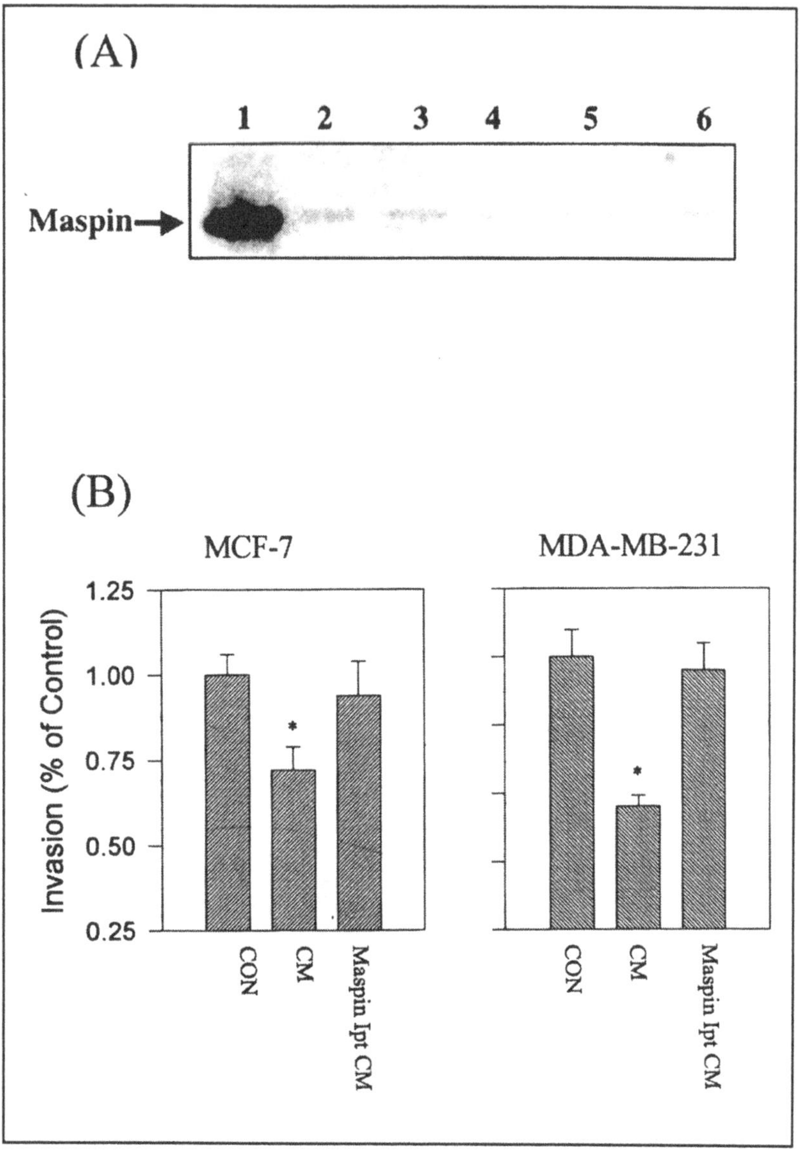

Fig. 2F. Maspin immunoprecipitation fraction [A] at various dilutions of maspin antibody (1/100), lane 1; (1/500), lane 2; (1/1000), lane 3; (1/2000), lane 4. Optimal dilution was 1/100 to achieve nearly 100% immunoprecipitation. Other serpin antibodies used including anti-PAI-1 (lane 5) and anti-PAI-2 (lane 6) resulted in only negligible cross-reacting immunoprecipitation of maspin. [B] Effects of HMS-1 25X CM and maspin-immunoprecipitated CM on breast carcinoma invasion. Control levels of invasion of designated breast carcinoma cell lines, MCF-7 and MDA-MB-231 were assigned arbitrary values of 1.0 and effects of CM and immunoprecipitated CM were expressed relative to these control levels. Results with other myoepithelial cell lines were similar. *indicates statistically significant differences compared to control ($p<0.05$).

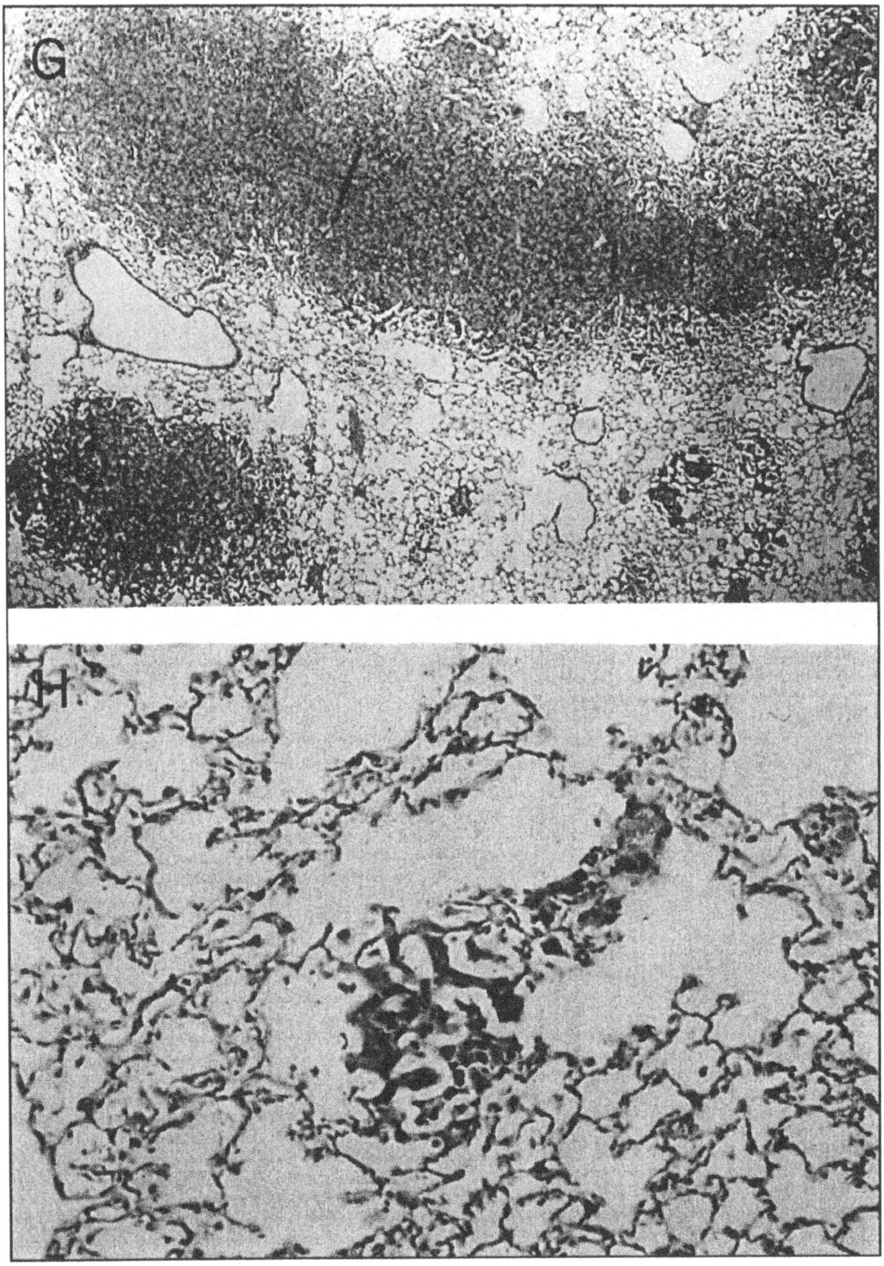

Fig. 2G & H. Differences in hematogenous pulmonary metastases with tail vein injected *neo*C8161 is in evidence in mice harboring non-myoepithelial xenografts (G) *v* myoepithelial xenografts (H). The number and size of metastatic colonies in mid-longitudinal cross section of lung was determined by digital image analysis and expressed as mean ± standard error.

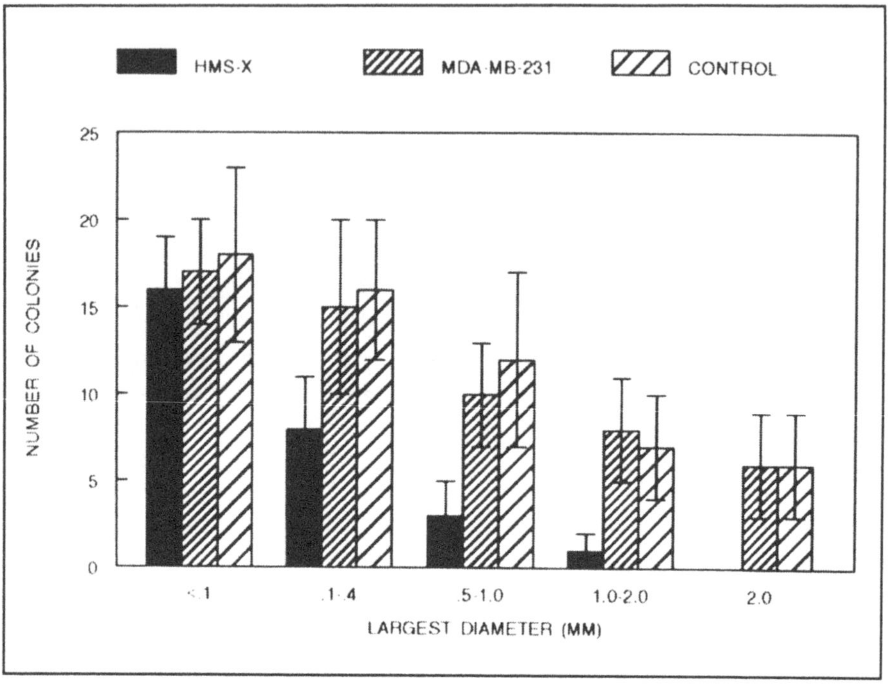

Fig. 21. Quantitation of pulmonary metastases revealed similar numbers of colonies in all three groups but a marked decrease in size in the group harboring the myoepithelial xenografts. Results depict a representative myoepithelial xenograft, HMS-X, a representative non-myoepithelial xenograft, MDA-MB-231, and control (no xenograft). Other myoepithelial and non-myoepithelial xenografts recapitulated these results.

FCS, or bFGF. When mixed with UVE, HMS-1 cells reduced endothelial migration to 12%±6% of control ($p<0.01$). HMS-1 concentrated CM reduced migration to 8%±7% of control ($p<0.01$). All of the non-myoepithelial malignant human cell lines studied stimulated both endothelial migration and proliferation. Concentrated CM from HMS-1, when fractionated on a heparin-Sepharose column, inhibited endothelial proliferation to 47%±10% of control ($p<0.01$). This inhibitory activity was present only in the 1.5-2.0M gradient fraction. Pretreatment of HMS-1 cells with PMA resulted in a 2-5 fold increase in endothelial antiproliferative inhibitory activity in both unfractionated CM as well as in the heparin-Sepharose fraction. Western blot of the heparin-Sepharose column fractions revealed that the 1.5-2.0M NaCl fraction contained thrombospondin-1. Immunoprecipitation of this fraction with anti-thrombospondin was effective at removing all thrombospondin-1 but decreased endothelial antiproliferative activity by only 50% raising the possibility that other angiogenic inhibitors including maspin were present in this fraction. The other myoepithelial cell lines (HMS-2-6) exhibited similar anti-angiogenic inhibitory activity in their fractionated and unfractionated CM. Therefore it is likely that both maspin and thrombospondin-1 are anti-angiogenesis effector molecules of myoepithelial cells.

To further explain our in vivo observations of minimal angiogenesis in our myoepithelial xenografts, in vitro and in vivo hypoxia studies were carried out. Non-myoepithelial xenografts, e.g., MDA-MB-231 exhibited florid hypoxia but only minimal necrosis when they reached a size of 2.0 cm. In contrast, the myoepithelial xenografts exhibited only minimal hypoxia but prominent necrosis ($p<0.001$) at the same size of 2.0 cm. Quantitation of the areas of hypoxia (pimonidazole immunoreactivity) and areas of necrosis in the myoepithelial *v* non-myoepithelial xenografts suggested that in the myoepithelial tumors where angiogenesis is minimal hypoxic areas progress to necrosis rapidly whereas in the non-myoepithelial tumors hypoxic areas accumulate but do not progress to necrosis presumably from the angiogenesis which the hypoxia elicits. Comparative analysis of myoepithelial *v* non-myoepithelial cell lines to low O_2 tension revealed that while both cell lines sense hypoxia in that they responded by increasing HIF-1α, the myoepithelial lines upregulated their steady state mRNA levels of the downstream genes, VEGF and iNOS to a lesser extent than the carcinoma lines suggesting the possibility of decreased transactivation of HRE. Specifically we observed an approximate 1.7-fold increase in VEGF (1.1-fold increase in iNOS) in myoepithelial cells in response to hypoxia compared to an approximate 2.5-fold increase in VEGF (1.5-fold increase in iNOS) in carcinoma cell lines in response to hypoxia. Although these fold differences by themselves were not impressive, the absolute levels of VEGF (and iNOS) expressed in carcinoma cells in response to hypoxia were 2.5-fold greater for VEGF (and 1.7-fold greater for iNOS) than the levels of VEGF (and iNOS) expressed in myoepithelial cells in response to hypoxia. Therefore it can be concluded that myoepithelial cells did not express VEGF or iNOS in response to hypoxia to nearly the same extent as carcinoma cells. To study both local and systemic effects of myoepithelial cells on metastasis, spontaneously metastasizing tumor cells were injected into our myoepithelial xenografts. The highly metastatic *neo*C8161 cells injected into the myoepithelial xenografts could be recovered in significant numbers although the numbers of clones recovered were less than those recovered from the nonmyoepithelial xenografts. Histological analysis of the extirpated xenografts revealed *neo*C8161 cells actively invading through all of the nonmyoepithelial xenografts in contrast to the appearance the myoepithelial xenografts where the *neo*C8161 cells were confined to the immediate areas around the injection site. Pulmonary metastases of *neo*C8161 were completely absent in the myoepithelial xenograft-injected group whereas they were quite numerous in the nonmyoepithelial group ($p<0.001$). Analysis of extirpated myoepithelial xenografts containing injected *neo*C8161 cells contained no evidence of murine angiogenesis by either vWf immunocytochemical studies or murine DNA Cot-1 analysis whereas a similar analysis of extirpated *neo*C8161 injected-nonmyoepithelial xenografts showed an increase in murine angiogenesis by both methods. This suggested that either the matrices of our myoepithelial xenografts or gene product(s) of the myoepithelial cells or both were inhibiting *neo*C8161-induced angiogenesis in vivo. We, in fact, found evidence of maspin, thrombospondin-1, TIMP-1, soluble bFGF receptors, prolactin and plasminogen fragments within 2M urea extracts of our myoepithelial xenografts. In tail vein injection studies of *neo*C8161, in mice harboring the myoepithelial xenografts, *neo*C8161 formed smaller pulmonary colonies than in mice harboring non-myoepithelial xenografts or in control mice (no xenografts) ($p<0.01$) (Fig. 2G, 2H, 2I). In a vWf factor immunocytochemical analysis of these smaller colonies in the mice harboring the myoepithelial xenografts, angiogenesis was minimal. These latter studies suggest the presence of circulating angiogenesis inhibitors released by the myoepithelial xenografts. Just recently we have demonstrated circulating maspin in mice harboring myoepithelial xenografts (see below).

Maspin and Myoepithelial Cells Can Be Manipulated Physiologically

Since PMA and dexamethasone were effective at pharmacologically altering maspin levels and the myoepithelial phenotype, we wondered whether physiological agents could do so as well. Because previous basic and clinical studies had examined the role of estrogen agonists and antagonists on human breast cancer cells and because issues of hormone replacement therapy (HRT) and tamoxifen chemoprevention are such timely issues in breast cancer, we wondered whether or not hormonal manipulations might affect myoepithelial cells in vitro as far as their paracrine suppressive activities on breast cancer were concerned. We recently demonstrated[12] that treatment of myoepithelial cells with tamoxifen but not 17-β estradiol increases both maspin secretion and invasion-blocking ability. 17-β Estradiol however competes with these suppressive effects of tamoxifen suggesting that the mechanism of tamoxifen action is estrogen receptor mediated. Myoepithelial cells lack ER-α but express ER-β. Tamoxifen, but not 17-β estradiol, increases AP-1 CAT but not ERE-CAT activity. Again, 17-β estradiol competes with the transcription-activating effects of tamoxifen. These experiments collectively suggest that the actions of tamoxifen on the increased secretion of maspin by myoepithelial cells may be mediated through ER-β and the trans-activation of an ER-dependent AP-1 response element.

As mentioned previously, immunoprecipitation of maspin from HMS-1 CM (Fig. 2F) reversed the anti-invasive effects of myoepithelial CM on breast carcinoma cell invasion in vitro. Tamoxifen treatment of HMS-1 resulted in a 2-3-fold increase in maspin secretion with increasing doses of tamoxifen and increasing times of exposure (Fig. 3A). 17-β Estradiol, in contrast, exerted no effects on maspin secretion and completely abolished the maspin stimulatory effects of tamoxifen in competition experiments. Tamoxifen's increase in maspin secretion was not due to an increase in steady state maspin mRNA levels which were essentially unchanged by this treatment. Myoepithelial cell lines lacked ER-α expression (Fig. 3B) but uniformly expressed ER-β (Figure 3B). Because the action of estrogen agonists/antagonists bound to estrogen receptors (either ER-α or ER-β) activates downstream genes containing either a classical ERE or an ER-dependent AP-1 response element, myoepithelial cell lines were transfected with CAT-reporter constructs fused to heterologous promoters containing the human estrogen response element (ERE-tk-CAT) or AP-1-tk-CAT. Tamoxifen (10^{-7} M) increased AP-1-CAT activity 3 fold (Fig. 3C). This effect was not observed with 17-β estradiol. Furthermore 17-β estradiol (10^{-5} M) competed with and blocked the effects of tamoxifen (10^{-7} M) (Fig. 3C). 17-β estradiol (10^{-7} M) did increase ERE-CAT activity but tamoxifen (10^{-7} M) did not.

Maspin and Myoepithelial Cells Are Potential Surrogate End Point Markers and Tumor Markers

Since maspin and myoepithelial cells seem intimately associated, and since myoepithelial cells are ubiquitous components of the ductal-lobular units of the breast and other organs which exhibit branching morphogenesis, we hypothesized that maspin might be detectable in fluid secreted by these ductal-lobular units. Since there has been a lot of recent interest in breast ductal fluid and breast nipple aspirates, especially, we measured maspin by Western blot and found it to be present in both nipple aspirates

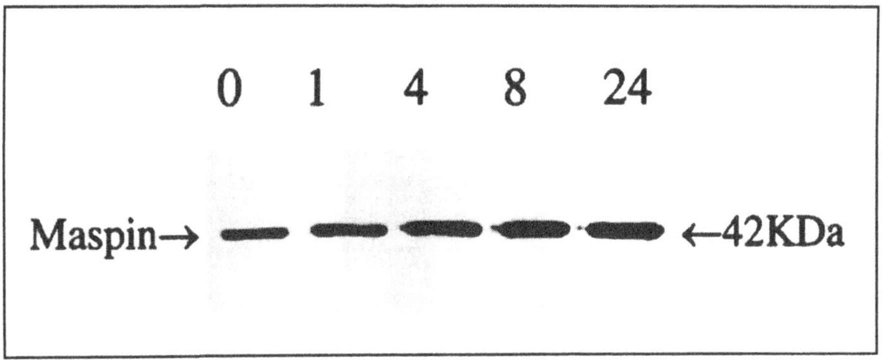

Fig. 3A. Effects of tamoxifen (10^{-7} M) treatment on maspin secretion for various times (hr) of tamoxifen exposure. 17β-Estradiol exerted no stimulatory effects. In competition experiments, increasing concentrations of 17β-estradiol completely blocked tamoxifen's stimulation of maspin secretion.

and ductal fluid (Fig. 4A) (Unpublished observations). This observation indicates that ductal fluid is not a mere transudate of blood or serum and that it is not a product of only epithelial cells (although epithelial protein products such as casein, lactalbumin and carcinoembryonic antigen (CEA) are certainly present). Ductal fluid also reflects a significant contribution from myoepithelial cells. From this observation we are currently studying groups of patients to see if their maspin levels serve to stratify them. We are currently analyzing ductal fluid collected following cannulation and washing of selected ducts in patients with microcalcifications on screening mammography who are about to undergo either excisional or core biopsy. Paired comparisons of maspin levels in ductal fluid obtained from ducts harboring microcalcifications or DCIS and normal ducts from the same patients are also being made. Maspin levels can be correlated with the histopathology surrounding the microcalcifications. It is anticipated that some of these patients will exhibit normal ductal histopathology surrounding their microcalcifications, some will harbor proliferations like hyperplasia, adenosis, ADH, and DCIS and still others invasive carcinoma. The screening value of maspin levels in all of these patients can be determined. Measurements of myoepithelial maspin in ductal fluid will be compared to levels of a breast epithelial cell marker such as CEA. CEA has been observed to be increased in nipple secretions and ductal fluid in patients with ductal hyperplasia. Hence the maspin/CEA ratio might be predictive of risk with increased maspin/CEA correlating with normalcy and decreased maspin/CEA correlating with either high risk, microcalcifications and/or precancerous histopathology. In this sense maspin can be used as a surrogate end point marker to predict either the risk of DCIS or the likelihood that DCIS will progress to invasive breast cancer.

Another interesting observation with respect to the use of myoepithelial maspin as a marker, this time, a tumor marker, is the observation that maspin can be detected in normal saliva but that it is markedly elevated in saliva secreted from a salivary gland neoplasm and that it is also elevated in murine serum in mice harboring human myoepithelial xenografts (Fig.4B) (unpublished observations). Most salivary gland neoplasms are thought to be myoepithelial in origin. These include mixed tumors, basal

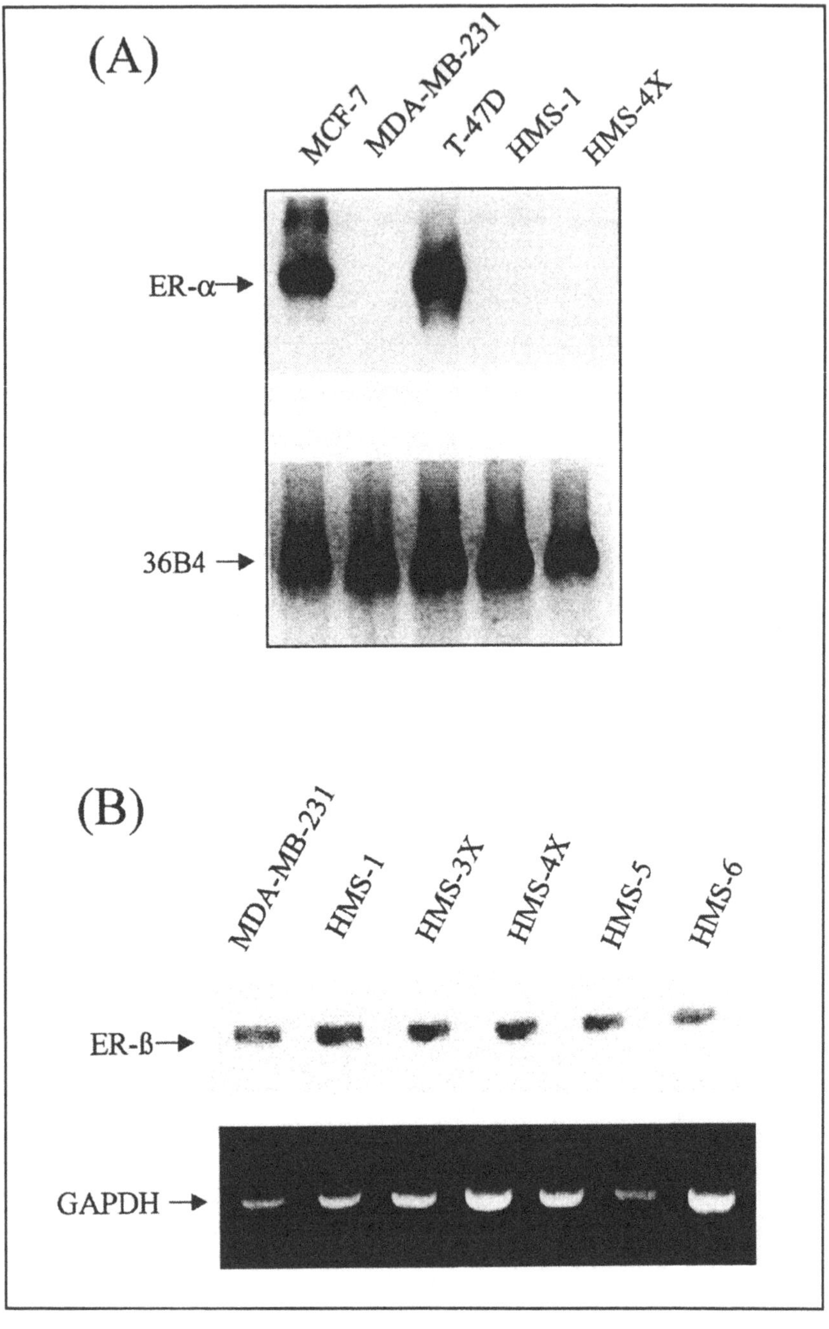

Fig. 3B. Expression of ER-α and ER-β in representative myoepithelial cell lines. [A] Northern blot analysis, ER-α expression. Normalization was with 36B4. [B] ER-β expression by RT-PCR. GAPDH served as a housekeeping control.

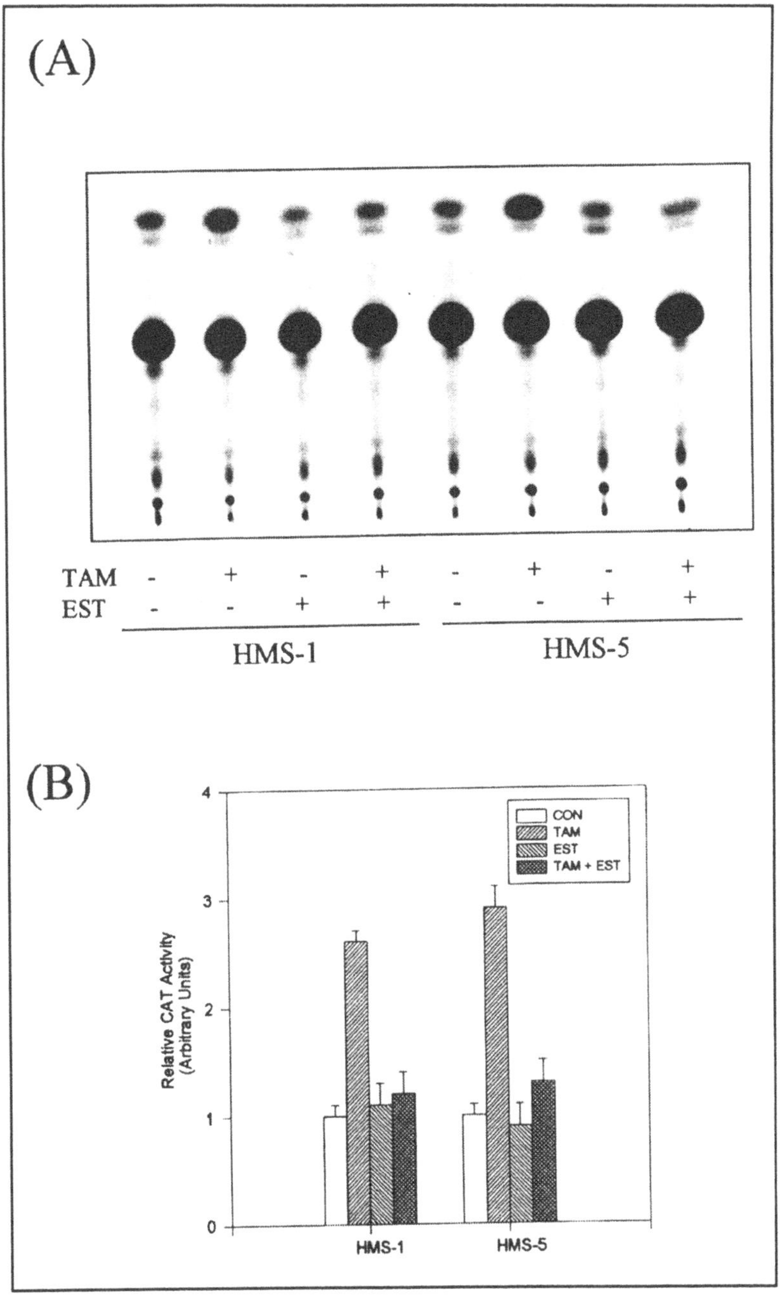

Fig. 3C. Tamoxifen (10^{-7} M) stimulation of AP-1-CAT activity in representative myoepithelial cell lines, HMS-1 and HMS-5 [A]. 17-β Estradiol (10^{-7} M) exerted no such stimulatory effects and blocked the effects of tamoxifen at high doses (10^{-5} M). [B] Results depicted are the means of three independent experiments. Error bars represent standard errors

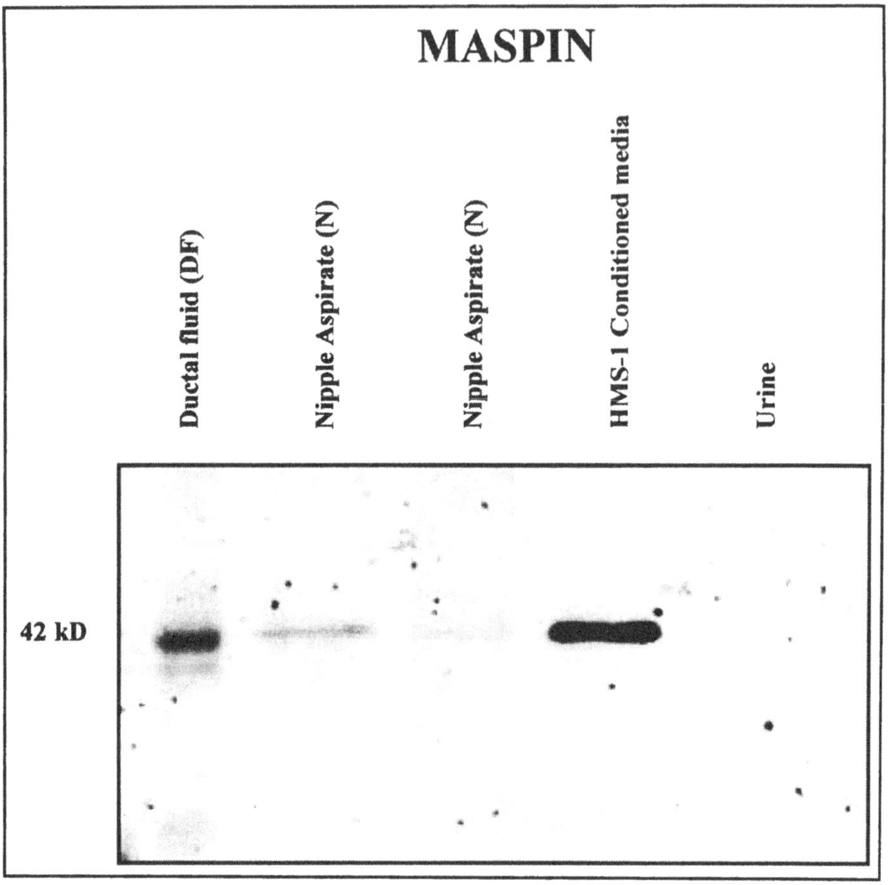

Fig. 4A. Nipple aspirate fluid (N) collected by nipple suction or ductal fluid (DF) collected by selected ductal cannulation can be studied for protein composition. In the former method (nipple aspiration), the contributions of each of the ductal systems can not be distinguished from each other; however in the latter method (ductal cannulation), each of the ductal systems can be separately analyzed and therefore ductal fluid from ducts harboring DCIS, for example, can be compared with ductal fluid from normal ducts in the same patient. Both nipple aspirate fluid and ductal fluid reflect the secretory contributions of the ductal-lobular units which are composed of both myoepithelial cells as well as epithelial cells. Maspin, a secretory product of myoepithelial cells, can be detected as a 42 kDa protein on Western blot from both nipple aspirate fluid (N) as well as ductal fluid (DF). Conditioned media (CM) from HMS-1, a myoepithelial cell line, is also intensely positive for maspin. Urine is negative.

cell adenomas, basal cell adenocarcinomas, and adenoid cystic carcinomas. It was human tumors of these types that originally gave rise to our myoepithelial cell lines/ xenografts that led to a dissection of the myoepithelial phenotype and to our observations concerning myoepithelial maspin. If screening saliva for maspin shows promise for detecting small incipient salivary gland neoplasms, then myoepithelial maspin will show its utility as a tumor marker. So, in summary, our findings indicate that maspin and myoepithelial cells contribute to the structural and functional integrity of the ductal-lobular units of different organs, and alterations in maspin levels in fluid from these units may reflect disease states (Fig.4C).

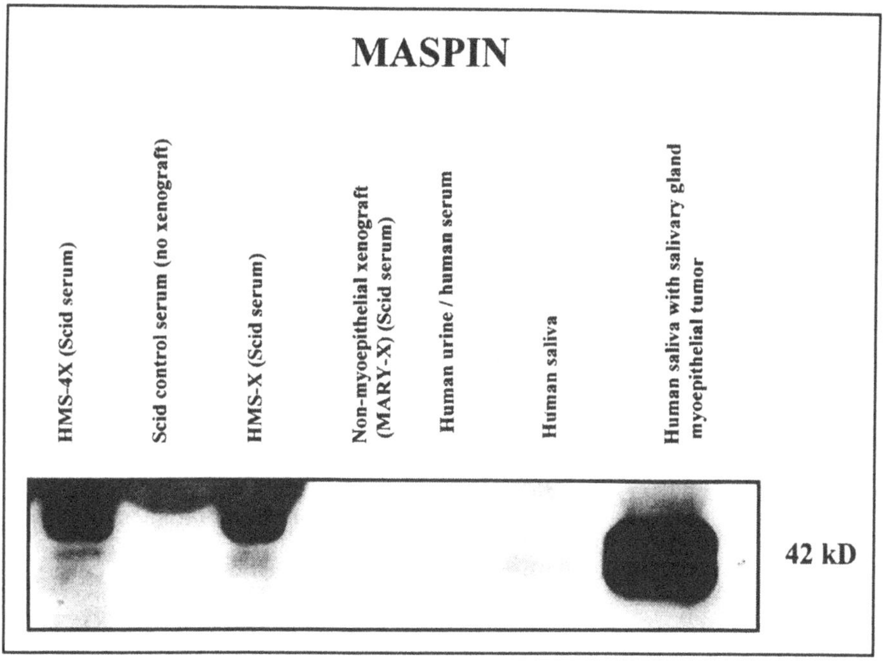

Fig. 4B. The contribution of myoepithelial tumors to maspin secretion in bodily fluids is demonstrated. In the Scid mouse, mice with >1 cm human myoepithelial xenografts (HMS-X, HMS-4X) exhibit circulating maspin levels in serum. Mice harboring no xenograft or a non-myoepithelial xenograft (MARY-X, an inflammatory breast carcinoma xenograft) contain no detectable circulating maspin. Human saliva in normal patients contain low levels of maspin, presumably from the myoepithelial cells in the salivary gland ductal-lobular units. Saliva from patients with benign salivary gland myoepithelial tumors show markedly elevated maspin.

Future Directions

The observations that myoepithelial cells secrete maspin in large quantities whereas carcinoma cells do not suggest that maspin and myoepithelial cells exert pleiotropic suppressive effects on tumor progression. Since maspin is both a proteinase inhibitor, a locomotion inhibitor and an angiogenesis inhibitor, the diverse actions of maspin may largely explain the pronounced anti-invasive and anti-angiogenic effects of myoepithelial cells on carcinoma and pre-carcinoma cells (13-24). Clearly, maspin and myoepithelial cells have more than marker value. Circulating maspin may have value as an anti-angiogenic agent. We need to better understand what it is about the myoepithelial phenotype that allows for high constitutive expression and secretion of maspin. Studies of the maspin promoter and *cis*/*trans* interactions within the myoepithelial cell seem to be an attractive line of further research. We also need to better understand the mechanism by which certain pharmacological agents such as PMA and certain physiological agents such as tamoxifen bolster myoepithelial secretion of maspin. With this understanding we may be able to design small less toxic molecules that have the same effect. We need to better exploit the intricate paracrine and local relationships which exist between myoepithelial cells and epithelial cells (precancerous and cancerous) in the breast and other organs. This is especially important and timely as

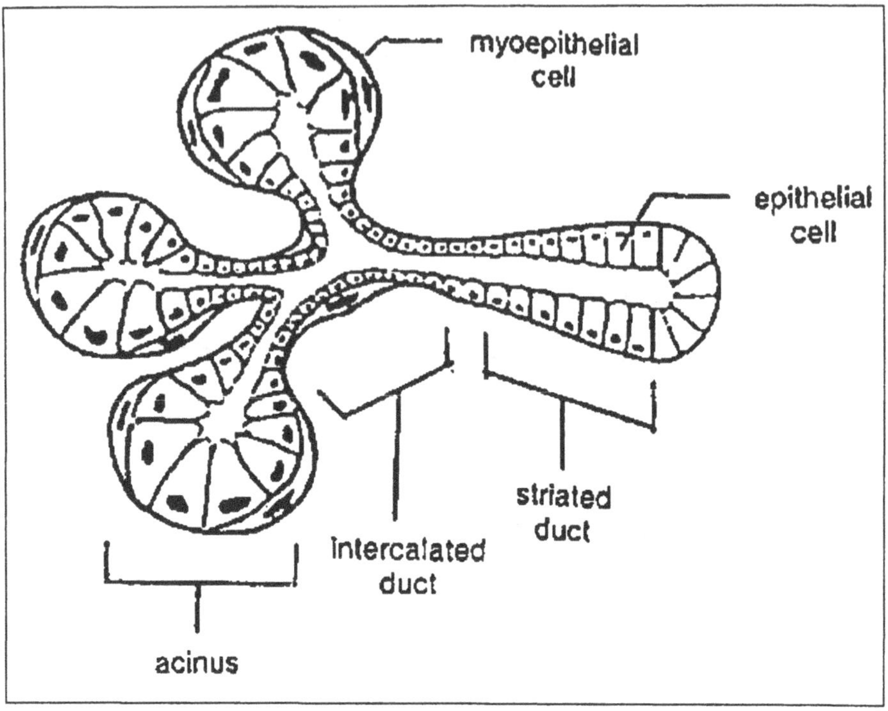

Fig. 4C. Schematic depicts the ductal-lobular unit of the breast and/or salivary gland and emphasizes the point that both myoepithelial cells as well as epithelial cells contribute to the integrity of this unit and to the composition of ductal fluid and saliva. In the case of invasive breast cancer, the integrity of the myoepithelial layer may be compromised and maspin levels may decrease. In the case of a myoepithelial tumor of salivary glands, the secretion of maspin into saliva may serve as a tumor marker.

intraductal approaches through the nipple are gaining in popularity as a means of screening women who are at risk for developing breast cancer. These intraductal approaches really exploit the local myoepithelial/epithelial relationships which exist. Screening for maspin levels as a surrogate end point marker is only the beginning. One could envision delivering intraductal gene therapy designed to exploit the inherent differences between myoepithelial and epithelial cells. One could target and destroy the epithelial cells selectively sparing the myoepithelium or alternately target the myoepithelial cells with a vector which bolsters its secretion of maspin. If the myoepithelial defense can be bolstered in this manner, perhaps this natural barrier which normally inhibits invasion for years can be made into an impervious barrier which inhibits invasion forever. At least that is one vision of scientists who are interested in maspin and myoepithelial cells.

Abbreviations

CM, conditioned medium; FCS, fetal calf serum; DCIS, ductal carcinoma in situ; PN-II/APP, protease nexin II/β-amyloid precursor protein; MMP, matrix metalloproteinase; TIMP, tissue inhibitor of metalloproteinase; uPA, urokinase-type plasminogen activator; PAI, plasminogen activator inhibitor; α1-AT, α1-antitrypsin; HMEC, human mammary epithelial cells; K-SFM, keratinocyte serum-free medium; vWf, von Willebrand factor; CHX, cyclohexamide; dB-cAMP, N^6,2'-O-dibutyryladenosine 3':5'-cyclic monophosphate; Na-But, sodium butyrate; RA, all *trans* retinoic acid; 5-azaC, 5-azacytidine; PMA, phorbol 12-myristate 13-acetate: UVE, umbilical vein endothelial cells; bFGF, basic fibroblast growth factor.

References

1. Cavenee WK. A siren song from tumor cells. J Clin Invest 1993; 91:3.
2. Liotta LA, Steeg PS, Stetler-Stevenson WG. Cancer metastasis and angiogenesis: An imbalance of positive and negative regulation. Cell 1991; 64:327-336.
3. Safarians S, Sternlicht MD, Freiman CJ, Huaman JA, Barsky SH. The primary tumor is the primary source of metastasis in a human melanoma/SCID model: Implications for the direct autocrine and paracrine epigenetic regulation of the metastatic process. Int J Cancer 1996; 66:151-158.
4. Folkman J, Klagsbrun M. Angiogenic factors. Science 1987; 235:442-447.
5. Cornil I, Theodorescu D, Man S, Herlyn M, Jambrosic J, Kerbel RS. Fibroblast cell interactions with human melanoma cells affect tumor cell growth as a function of tumor progression. Proc Natl Acad Sci USA 1991; 88:6028-6032.
6. Guelstein VI, Tchypsheva TA, Ermilova VD, Ljubimov AV. Myoepithelial and basement membrane antigens in benign and malignant human breast tumors. Int J Cancer 1993; 53:269-277.
7. Cutler LS. The role of extracellular matrix in the morhphogenesis and differentiation of salivary glands. Adv Dent Res 1990; 4:27-33.
8. Sternlicht MD, Safarians S, Calcaterra TC, Barsky SH. Establishment and characterization of a novel human myoepithelial cell line and matrix-producing xenograft from a parotid basal cell adenocarcinoma. In Vitro Cell Dev Biol 1996; 32:550-563.
9. Sternlicht MD, Kedeshian P, Shao ZM, Safarians S, Barsky SH. The human myoepithelial cell is a natural tumor suppressor. Characterizations of the extracellular matrix and proteinase inhibitor content of human myoepithelial tumors. Clin Cancer Res 1997; 3:1949-1958.
10. Nguyen M, Lee MC, Wang JL, Tomlinson JS, Shao ZM, Barsky SH. The human myoepithelial cell displays a multifaceted anti-angiogenic phenotype. Oncogene 2000; 19:3449-3459.
11. Zhang M, Volpert O, Shi YH, Bouck N. Maspin is anangiogenesis inhibitor. Nature Med 2000; 6:196-199.
12. Shao ZM, Radziszewski WJ, Barsky SH. Tamoxifen enhances myoepithelial cell suppression of human breast carcinoma progression by two different effector mechanisms. Cancer Lett 2000; 157:133-144.
13. Andreasen PA, Georg B, Lund LR, Riccio A, Stacey SN. Plasminogen activator inhibitors: Hormonally regulated serpins. Mol Cell Endocrinol 1990; 68:1-19.
14. Albini A, Iwamoto Y, Kleinman HK, Martin GS, Aaronson SA, Kozlowski JM et al. A rapid *in vitro* assay for quantitating the invasive potential of tumor cells. Cancer Res 1987; 47:3239-3245.
15. Kataoka H, Seguchi K, Iwamura T, Moriyama T, Nabeshima K, Koono M. Reverse-zymographic analysis of protease nexin-II/amyloid b protein precursor of human carcinoma cell lines, with special reference to the grade of differentiation and metastatic phenotype. Int J Cancer 1995; 60:123-128.
16. Narindrasorasak S, Lowery DE, Altman RA, Gonzalez-DeWhitt PA, Greenberg BD, Kisilevsky R. Characterization of high affinity binding between laminin and Alzheimer's disease amyloid precursor proteins. Lab Invest 1992; 67:643-652.
17. Rao CN, Liu YY, Peavey CL, Woodley DT. Novel extracellular matrix-associated serine proteinase inhibitors from human skin fibroblasts. Arch Biochem Biophys 1995; 317:311-314.

18. Zou Z, Anisowicz A, Hendrix MJC, Thor A, Neveu M, Sheng S et al. Maspin, a serpin with tumor-suppressing activity in human mammary epithelial cells. Science 1994; 263:526-529.
19. Hopkins PCR, Whisstock J. Function of maspin. Science 1994; 265:1893-1894.
20. Pemberton PA, Wong DT, Gibson HL, Kiefer MC, Fitzpatrick PA, Sager R et al. The tumor suppressor maspin does not undergo the stressed to relaxed transition or inhibit trypsin-like serine proteases: Evidence that maspin is not a protease inhibitory serpin. J Biol Chem 1995; 270:15832-15837.
21. Sheng S, Pemberton PA, Sager R. Production, purification, and characterization of recombinant maspin proteins. J Biol Chem 1994; 269:30988-30993.
22. Folkman J. Clinical applications of research on angiogenesis. N Engl J Med 1995; 333:1757-1763.
23. Roberts DD. Regulation of tumor growth and metastasis by thrombospondin-1. FASEB J 1996; 10:1183-1191.
24. Weinstat-Saslow DL, Zabrenetzky VS, VanHoutte K, Frazier WA, Roberts DD, Steeg PS. Transfection of thrombospondin 1 complementary DNA into a human breast carcinoma cell line reduces primary tumor growth, metastatic potential, and angiogenesis. Cancer Res 1994; 54:6504-6511.

Chapter 5

Maspin and Pericellular Plasminogen Activation in Cell-Matrix Interaction

Shijie Sheng, Hector Biliran Jr. and Richard McGowen

Introduction

Maspin may offer a unique opportunity to block tumor invasion and metastasis. Maspin expression correlates with normality, and pre-malignant and/or less invasive lesions in breast, prostate and oral squamous cells.[1-5] Therefore, maspin may be a useful molecular marker for the diagnosis and/or prognosis of cancer. Furthermore, accumulated evidence supports a tumor suppressive role of maspin, at the level of invasion and metastasis. For example, orthotopic explants of mammary carcinoma cells transfected with maspin coding cDNA are inhibited in tumor growth and metastasis in nude mouse experiments.[1] These maspin transfectants are significantly inhibited in in vitro invasion and motility assays.[1,6] Consistently, three forms of recombinant human maspin proteins produced in the bacterium *E. coli*, yeast *S. cerevisiae*, and baculo-virus infected insect cells *Sf9*, respectively, exhibit potent inhibitory effects on the invasion and motility of an array of human mammary and prostatic carcinoma cell lines in vitro.[7,8] Treatment of human mammary carcinoma cells with maspin leads to a partial restoration of benign epithelial morphology and an increased cell adhesion to fibronectin.[9] Recently, Zhang et al showed that recombinant mouse maspin inhibits the metastasis of human prostate tumor xenografts in nude mice.[10] These data suggest a potential clinical application of maspin in cancer intervention.

To develop a maspin-based anti-cancer therapy, it is essential to understand the molecular mechanism of maspin. Based on the protein sequence deduced from the maspin cDNA, maspin protein has an overall homology to serine protease inhibitors, or serpins. The following Chapter is focused on the novel biochemical properties of maspin as well as its complex role in pericellular proteolysis and cell-matrix interaction.

Maspin on Soluble Serine Proteases

Maspin is closely related to several serpins including: equine and human neutrophil-monocyte elastase inhibitor,[1] human squamous carcinoma antigens 1 and 2,[11] plasminogen activator inhibitor type 2,[1] protease inhibitor-8,[12] protease inhibitor-10[12] and chick ovalbumin.[1] However, in vitro biochemical studies as summarized in Table 1 show that purified recombinant maspin does not inhibit several purified serine

Maspin, edited by Mary J.C. Hendrix. ©2002 Eurekah.com

Table 1. The effect of recombinant maspin on serine proteases

Protease	Free [1]	Co-coated with Fibrinoen[2]	On Cell Surface[3]	In CM[4]
uPA	-	-	Ki = 20 nM	-
tPA	Stimulatory	Ki = 130 nM	NA	-
Plasmin	Substrate	-	-	-
Chymotrypsin	-	-	ND	ND
Trypsin	Substrate	-	ND	ND
Elastase	-	-	ND	ND
Thrombin	-	-	ND	ND

-, No detectable effect;
NA, The corresponding target enzyme was not detected;
ND, Not determined;
1, Assayed under the optimal condition for each corresponding enzyme in the absence of co-factors;
2, Assayed under the optimal condition for tPA in the presence of fibrinogen and gelatin;
3, Assayed with intact monolayer DU145 cells;
4, Assayed with purified enzyme in the serum-free conditioned medium of DU145 Cells.

proteases such as urokinase-plasminogen activator (uPA), trypsin, chymotrypsin, elastase, plasmin, and thrombin.[13,14]

It is known that a serpin uses its pseudosubstrate-like peptide bond (p_1-p_1') in a reactive site loop (RSL) to interact with the catalytic site of the target serine protease. Maspin with an Arginine aligned at its p_1' position is predicted to specifically interact with trypsin-like proteases.[1] Interestingly, when incubated with purified trypsin or plasmin, maspin was specifically cleaved at the predicted p_1-p_1' peptide bond.[7,13] These results indicate a specific interaction between maspin RSL and the catalytic sites of those proteases, albeit the kinetics of such interaction favors the cleavage of maspin in solution.

A current hypothesis suggests that upon initial docking of the target protease, an inhibitory serpin undergoes dramatic conformational changes and eventually forms a stable 1:1 stoichiometric complex with the protease. Prior to this inhibitory interaction, a partial pre-insertion of RSL into the β-pleated sheet structure (stressed conformation, S) has been noted in several serpins.[15-18] One exception of this general phenomenon is inhibitory α1-antichymotrypsin, whose crystal structure reveals a fully exposed RSL.[19,20] Thus, it remains unclear whether RSL pre-insertion is a prerequisite for all inhibitory serpins. Nonetheless, a school of thought predicted that the novel RSL sequence of maspin (aa 320-340) might render it non-inhibitory.[13,21] Based on crystallographic analyses of several inhibitory serpins, a highly conserved hinge region located 9-15 residues amino terminal to the reactive site p_1-p_1' bond is critical for the partial RSL pre-insertion, as well as for the subsequent formation of the protease/inhibitor complex.[22,23] The hinge sequence of maspin is somewhat deviant from and 4 residues shorter than that of a classical inhibitory serpin.[1,24] Homology modeling using either non-inhibitory serpin chick ovalbumin[21] or α1-antichymotrypsin as a template reveals that maspin RSL is an exposed, but not pre-inserted loop (Fig. 1). Consistent with this model, recombinant maspin does not have a stressed conformation in solution. Furthermore, heat treatment that is known to induce the transition of classical inhibitory serpins from the stressed conformation (S) to the relaxed conformation (R, with fully inserted RSL) does not induce the S-R transition in maspin.[13]

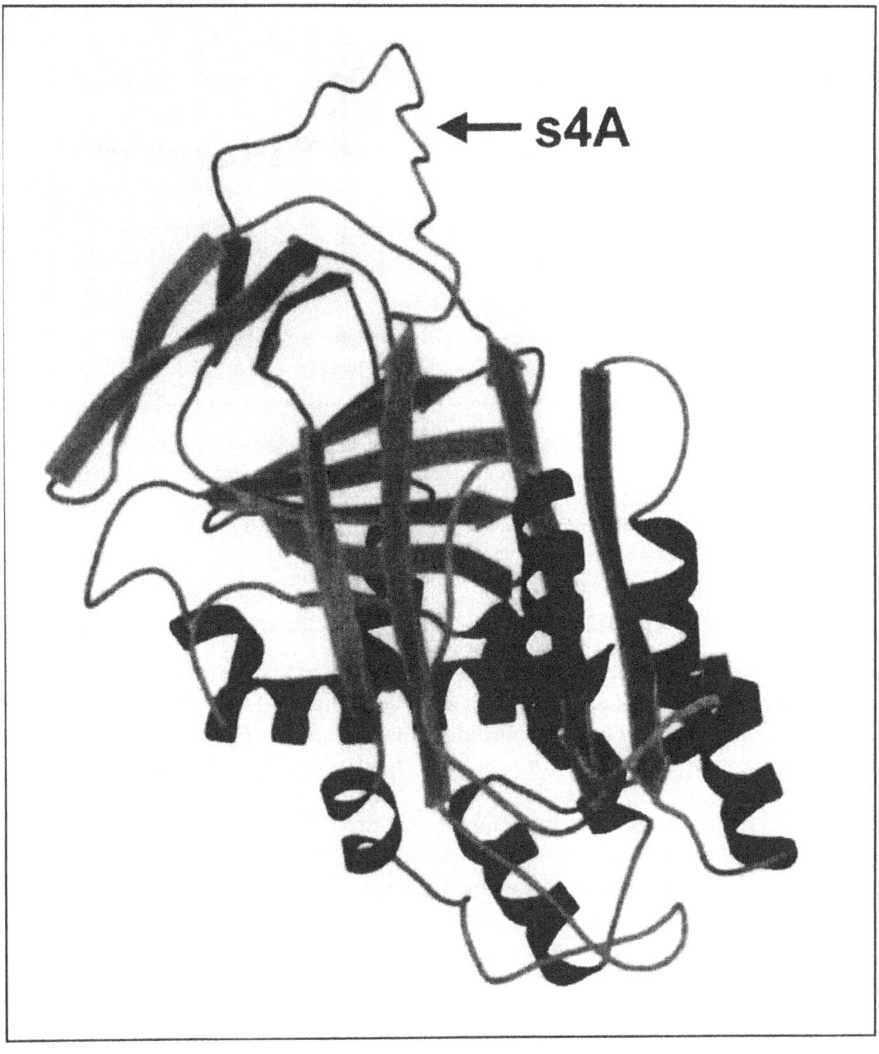

Fig. 1. Homology modeling of maspin using α1-antichymotrypsin as a template.

However, maspin has been shown to undergo conformational changes under several other conditions. For example, purified monomeric recombinant maspin undergoes spontaneous three-state unfolding and polymerization under cell-free conditions.[7,13,25] According to detailed x-ray crystallographic studies of several classical inhibitory serpins, serpin dimerization and polymerization involves the insertion of RSL into the β-pleated sheets.[26,27] Interestingly, the dimerization of maspin during long-term storage is concomitant with specific cleavage at the p_1-p_1' site. Furthermore, both the dimerized and cleaved maspin exhibit a higher affinity for the heparin column as compared to the intact monomeric maspin (Fig. 2), indicating major conformational changes not only in the RSL but also in the heparin-binding domain. Thus, despite its non-inhibitory activity in solution, maspin appears to be capable of undergoing major conformational changes such as RSL insertion.

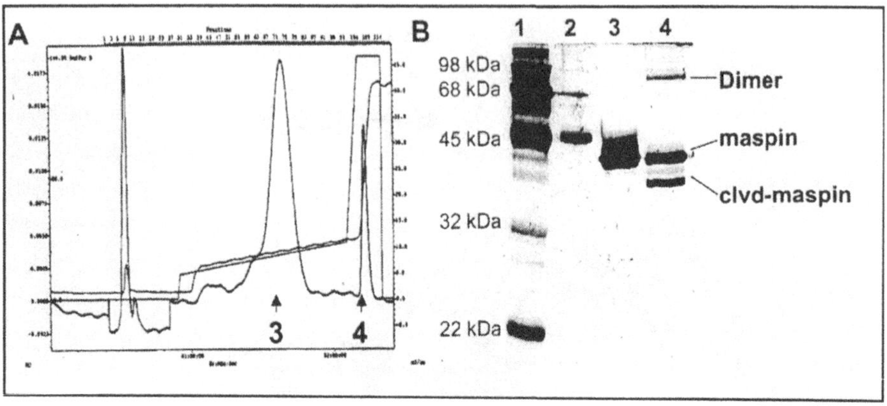

Fig. 2. HPLC purification of monomeric maspin from polymerized and cleaved maspin using a heparin affinity column. (A): the elution chromatogram showing the salt gradient and protein absorbance at 280 nm; (B) Silver-stained SDS-PAGE of purified proteins. Lanes 1-4 are protein molecular weight standards, purified chick ovalbumin, elution fraction from peak number 3, and elution fraction from peak number 4, respectively.

Maspin with Bound Plasminogen Activators

In light of the documented structural flexibility of serpins and the common requirements of serpin co-factors for their inhibitory function,[22,28-30] current structural considerations do not exclude the possibility that specific biological microenvironment may foster a maspin conformation in favor of an enhanced inhibitory capacity. On the other hand, biological studies showed that the inhibitory effect of maspin on tumor cell invasion and motility requires the intact maspin RSL. An oligopeptide-derived antibody against maspin RSL blocks the inhibitory activity of maspin in both invasion and motility assays.[8] Limiting digestion of maspin by trypsin at the p_1-p_1' site abrogates the biological activity of maspin in invasion assay.[7] In addition, Sheng, et al discovered that immobilized maspin specifically bound single-chain tPA in the culture medium conditioned by breast carcinoma cells MDA-MB-435.[14] These data suggest that maspin may indeed act as an inhibitory serpin.

It is important to note that maspin acts on the cell membrane to inhibit tumor cell motility and invasion.[8,9] Endogenous maspin is present both as a soluble cytoplasmic protein, as well as associated with secretory vesicles and the plasma membrane.[7,31] The first 50 N-terminal amino acids of maspin have been identified responsible for its localization on cell membrane.[31] Under a nonpermeabilizing condition, maspin bound to the surface of prostate carcinoma cells DU145 in a saturable manner, suggesting that it may specifically interact with a cell surface-associated molecule.[32] It is also important to bear in mind that some serine proteases are tightly regulated by their associated molecules. For example, tPA is associated and activated by fibrin, fibrinogen and several other extracellular matrix proteins.[33-35] uPA, on the other hand, has been shown to be activated by its cell-surface receptor uPAR.[36-39] Thus it is important to investigate the biochemical activity of maspin under an in vivo-like condition.

The first evidence that maspin inhibits a serine protease was found by Sheng et al using tPA that is preactivated by fibrinogen.[14] Purified maspin exerts a biphasic effect on fibrinogen-associated tPA, acting as a competitive inhibitor (Ki = 130 nM) at low concentrations (< 0.5 μM) and as a stimulator at higher concentrations. Interestingly, maspin C-terminally truncated at p_1 position exhibits an exclusively stimulatory effect on either free tPA or fibrinogen-associated tPA. A more detailed kinetic study lead to a hypothesis that two segregated domains of maspin interacts with the catalytic and activating domain of tPA, respectively (Fig. 3). The dependence of the inhibitory activity of maspin on a tPA co-factor, fibrinogen, raises the possibility that maspin may become a stronger inhibitor if the activation domain of the target enzyme is blocked.

The antiproteolytic activity of maspin correlates well with its biological activity in inhibiting tumor invasion and metastasis. However, it is well documented that uPA (along with uPAR), but not tPA, is causatively involved in the invasive and metastatic phenotypes, and the poor prognosis of many types of carcinomas.[40-42] Thus, it is important to find out whether a potential inhibitory interaction between maspin and the tumor cell-associated uPA underlies the molecular mechanism of maspin. To this end, a recent report in by McGowen et al showed that the tumor cell surface-associated uPA was specifically inhibited by recombinant maspin.[32]

To address the specific interaction between maspin and the cell surface-associated uPA, McGowen and colleagues took advantage of the established prostatic carcinoma cell line DU145 that produces uPA as the predominant plasminogen activator, but does not express detectable level of endogenous maspin. It was shown that the inhibitory effect of maspin was similar to that of a specific uPA-neutralizing antibody, and was reversed by the polyclonal antibody made against the maspin RSL sequence. The essential role of the intact RSL of maspin in its proteolytic inhibitory activity towards cell-associated uPA further indicates that maspin acts as an inhibitory serpin. Detailed kinetic analyses revealed that maspin acted as a competitive inhibitor of uPA with an apparent Ki value of 20 nM. These kinetic characteristics are comparable to those of PAI-1 and PAI-2, respectively, in similar cell-based biochemical analyses.

Interestingly, unlike plasminogen activator inhibitors PAI-1 and PAI-2, maspin only inhibits the cell surface-associated uPA, but not the secreted uPA or purified uPA in cell-free biochemical reactions. In addition, maspin formed a stable complex only with the uPA in the cell lysate fraction of DU145 cells, but not with the purified uPA or the uPA secreted by DU145 cells into the conditioned culture medium.[32] These data demonstrated an important role of epithelial cell surface in mediating the inhibitory interaction between maspin and uPA. Since the activation and the activity of uPA on the cell surface are mediated by its receptor uPAR,[36-39] it is reasonable to postulate that the association with uPAR may render uPA prone to the inhibition by maspin. On the other hand, the cell surface microenvironment may also provide additional co-factor(s) that further increases the inhibitory potency of maspin by facilitating its critical transition from a latent conformation to an active conformation.

Maspin in Cell-Matrix Interaction

According to a three-step hypothesis that describes the sequential events of tumor–cell invasion of the extracellular matrix (ECM), tumor cells first attach to the matrix (adhesion) through a set of ECM proteins (such as laminin and fibronectin) and their transmembrane integrin receptors. After the attachment, tumor cells secret proteolytic enzymes that can locally degrade the matrix. Then tumor cells migrate into the resulting void region of the matrix.[43] Previous in vitro studies using both

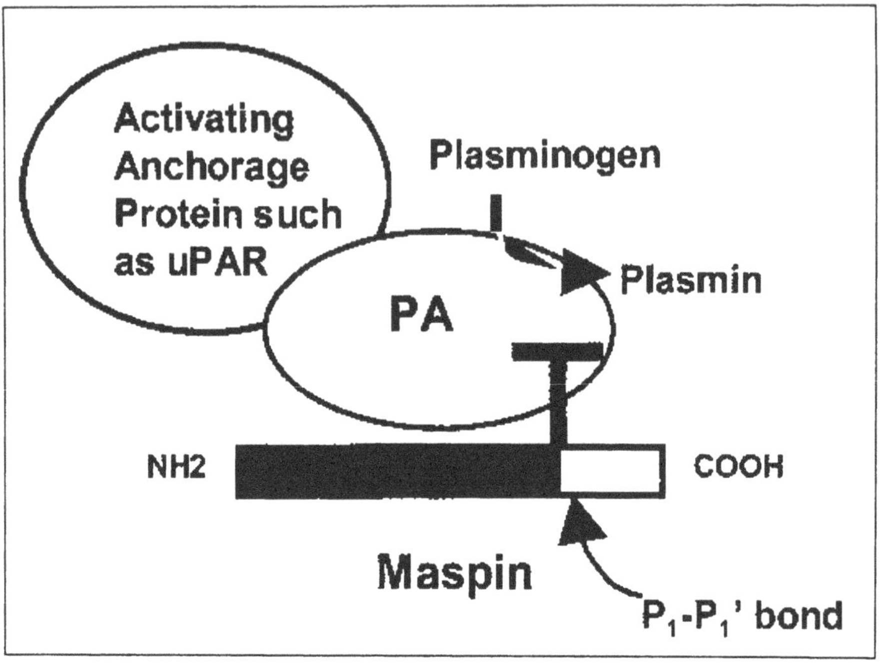

Fig. 3. Hypothetical model for interactions between maspin and plasminogen activator.

endogenously expressed maspin or purified recombinant maspin protein have demonstrated that the inhibitory effect of maspin on tumor invasion and metastasis is, at least in part, due to its localized action at the interphase between cell membrane and the extracellular matrix (ECM).[8] Figure 4 illustrates our current hypothesis on how maspin may exert its inhibitory effect on tumor cell motility and invasion.

The distinct membrane-dependent antiproteolytic activity of maspin quantitatively correlates with its biological activity in inhibiting tumor cell invasion and motility,[14,32] and suggests that maspin may directly inhibit ECM degradation. In this regard, the link between maspin and cell surface-associated uPA is of particular importance. By virtue of its proteolytic activity, elevated uPA may contribute to the catastrophic destruction of basement membrane (BM) and ECM. BM and ECM degradation may further induce cell detachment from the established adhesion, which in turn propels the cells to migrate.

In addition, as reported earlier by Seftor et al, maspin treatment of breast carcinoma cells MDA-MB-435 lead to an increased cell surface expression of $\alpha_3\beta_1$ integrin, which was correlated with an increased cell adhesion to fibronectin.[9] Although the experimental attachment of suspended cells to a extracellular substratum may not be identical to the attachment step during the concerted cell migration within ECM, the data of Seftor et al suggests an important role of maspin in the establishment of focal contacts, raising the possibility that a sustained reinforcement of cell adhesion by maspin may further prevent cell migration. To date, the underlying molecular mechanism for the maspin effect on cell adhesion is unclear. It is worth noting that the amino terminal fragment of uPA has been shown to bind to its glycosylphosphotidylinositol (GPI)-linked cell surface receptor uPAR and exert a complex stimulatory effect on the

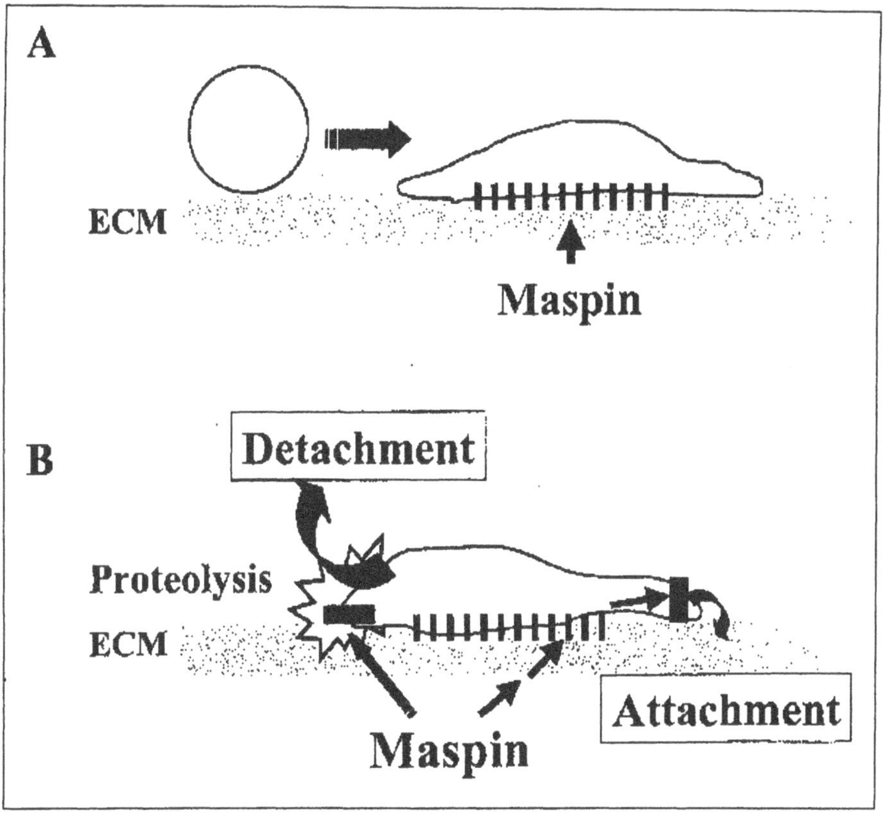

Fig. 4. Hypothetic model for the role of maspin in pericellular proteolysis and cell-matrix interaction. (A) A role of maspin in the initial attachment of suspended cells to a surface coated with ECM; (B) A role of maspin in the muti-step cell migration and invasion.

adhesion and migration of several types of cells.[36-39] In fact, increasing amount of evidence indicates that uPA/uPAR directly interacts with ECM receptor integrins.[44-47] It remains to be tested whether interaction of maspin with uPA/uPAR complex subsequently regulates the uPAR interaction with focal adhesion complex.

It is important to point out, that different plasminogen activator inhibitors may play distinct roles in tumor progression. For example, PAI-1 along with uPA and uPAR, is causatively involved in the progression of breast cancer.[48-50] In contrast, maspin, which is down-regulated in several types of carcinomas, has a tumor suppressive activity. Interestingly, PAI-1 is associated with poor prognosis of several types of cancers[48-50] and is associated with increased tumor metastasis in mice.[51] At the molecular level, PAI-1 has been shown to regulate the uPAR-mediated cell adhesion to vitronectin by directly binding to the sometomedin-B like domain of vitronectin.[52,53] This paradoxical role of PAI-1 lead to a better appreciation of the specialized and complex functionality of a serpin in regulating the activity of uPA/uPAR. It is reasonable to predict that the molecular interaction between maspin and uPA/uPAR may be regulated by additional structural determinants of maspin and by its associated molecules on the cell surface.

Conclusion and Perspectives

It is likely that maspin inhibiting tumor invasion by inhibiting uPA-mediated proteolysis at the interphase between cell membrane and extracellular matrix. Given the documented interaction between uPA/uPAR and cell focal adhesion complex,[44-46] the specific interaction between maspin and cell surface-associated uPA complex may also lead to changes of cell surface integrin profile in favor of stabilizing the established cell-matrix interaction, further preventing cell migration. Extensive studies by others have shown a causative involvement of uPA in the invasive and metastatic phenotype of many types of carcinomas.[40-42] Conversely, inhibition of uPA activity dramatically reduces tumor invasiveness.[54,55] While future studies are needed to test whether endogenous maspin protein has a similar proteolytic inhibitory effect on the cell surface-associated uPA, it is intriguing to hypothesize that novel maspin-based therapeutic strategies may prove useful to specifically target human malignancies that are associated with elevated uPA/uPAR.

Tumor metastasis consists of multiple steps including: overproliferation, invasion of basement membrane and stroma, angiogenesis, intravasation, escape of host immune surveillance, extravasation, and tumor growth at distal sites. Several lines of evidence also linked maspin to cell growth inhibition and apoptosis. Orthotopic explants of maspin transfectants derived from breast carcinoma cells are inhibited in tumor growth in nude mice.[1] In transgenic mice that overexpress mouse maspin under the control of the WAP promoter, mammary development is significantly blocked, possibly due to increased epithelial cell apoptosis.[56] Furthermore, induced expression of maspin in two breast carcinoma cell lines correlates with cell cycle arrest at G_2 phase induced by the $p21^{WAF1/CIP}$ pathway.[57] Recently, Zhang et al showed that local delivery of recombinant mouse maspin lead to apoptosis of prostate tumor xenografts in nude mice and inhibited tumor-induced neovascularization.[10] Since maspin protein, either endogenously re-expressed or exogenously added as purified recombinant protein, is not sufficient to inhibit the growth of several breast carcinoma cell lines in vitro,[7,58] it remains unclear whether maspin is a regulator or an effector of cell growth/apoptosis signals. Future studies are needed to address whether maspin regulates cell growth and apoptosis by a mechanism similar to that involved in cell motility and invasion.

Acknowledgements

Foremost, the authors wish to thank Dr. Arthur B. Pardee for his generous support. We wish to thank Dr. Bienkovska Jadviga. for her help in computer modeling of maspin. This work was supported in part by a Prostate cancer Pilot Grant form National Cancer Institute/Wayne State University CA69845 (to S. S.), a Ruth Sager Memorial Fund established by Dr. Arthur B. Pardee through Wayne State University (to S. S.), and National Institute of Health Grant CA84176 (to S. S.).

References

1. Zou Z, Anisowicz A, Hendrix MJ, Thor A, Neveu M, Sheng S et al. Maspin, a serpin with tumor-suppressing activity in human mammary epithelial cells. Science 1994; 263:526-529.
2. Xia W, Lau YK, Hu MC, Li L, Johnston DA, Sheng S et al. C. High tumoral maspin expression is associated with improved survival of patients with oral squamous cell carcinoma. Oncogene 2000; 19:2398-2403.

3. Barsky SH, Doberneck SA, Sternlicht MD, Grossman DA, Love SM. 'Revertant' DCIS in human axillary breast carcinoma metastases. J Pathol 1997; 183:188-194.
4. Luppi M, Morselli M, Bandieri E, Federico M, Marasca R, Barozzi P et al. Sensitive detection of circulating breast cancer cells by reverse-transcriptase polymerase chain reaction of maspin gene. Ann Oncol 1996; 7:619-624.
5. Zhang M, Magit D, Sager R. Expression of maspin in prostate cells is regulated by a positive ets element and a negative hormonal responsive element site recognized by androgen receptor. Proc Natl Acad Sci USA 1997; 94:5673-5678.
6. Zhang M, Maass N, Magit D, Sager R. Transactivation through ets and Ap1 transcription sites determines the expression of the tumor-suppressing gene maspin. Cell growth Differ 1997; 8:179-186.
7. Sheng S, Pemberton PA, Sager R. Production, purification, and characterization of recombinant maspin proteins. J Biol Chem 1994; 269:30988-30993.
8. Sheng S, Carey J, Seftor EA, Dias L, Hendrix MJ, Sager R. Maspin acts at the cell membrane to inhibit invasion and motility of mammary and prostatic cancer cells. Proc Natl Acad Sci USA 1996; 93:11669-11674.
9. Seftor RE, Seftor EA, Sheng S, Pemberton PA, Sager R, Hendrix MJ. maspin suppresses the invasive phenotype of human breast carcinoma. Cancer Res 1998; 58:5681-5685.
10. Zhang M, Volpert O, Shi YH, Bouck N. Maspin is an angiogenesis inhibitor. Nat Med 2000; 6:196-199.
11. Schneider SS, Schick C, Fish KE, Miller E, Pena JC, Treter SD et al. A serine proteinase inhibitor locus at 18q21.3 contains a tandem duplication of the human squamous cell carcinoma antigen gene. Proc Natl Acad Sci USA 1995; 92: 3147-3151.
12. Bartuski AJ, Kamachi Y, Schick C, Overhauser J, Silverman GA. Cytoplasmic antiproteinase 2 (PI8) and bomapin (PI10) map to the serpin cluster at 18q21.3. Genomics 1997; 43:321-328.
13. Pemberton PA, Wong DT, Gibson HL, Kiefer MC, Fitzpatrick PA, Sager R et al. The tumor suppressor maspin does not undergo the stressed to relaxed transition or inhibit trypsin-like serine proteases. Evidence that maspin is not a protease inhibitory serpin. J Biol Chem 1995; 270:15832-15837.
14. Sheng S, Truong B, Fredrickson D, Wu R, Pardee AB, Sager R. Tissue-type plasminogen activator is a target of the tumor suppressor gene maspin. Proc Natl Acad Sci USA 1998; 95:499-504.
15. Pemberton PA, Stein PE, Pepys MB, Potter JM, Carrell RW. Hormone binding globulins undergo serpin conformational change in inflammation. Nature 1988; 336:257-258.
16. Pemberton PA, Harrison RA, Lachmann PJ, Carrell RW. The structural basis for neutrophil inactivation of C1 inhibitor. Biochem J 1989; 258:193-198.
17. Mast AE, Enghild JJ, Pizzo SV, Salvesen G. Analysis of the plasma elimination kinetics and conformational stabilities of native, proteinase-complexed, and reactive site cleaved serpins: Comparison of alpha 1-proteinase inhibitor, alpha 1-antichymotrypsin, antithrombin III, alpha 2-antiplasmin, angiotensinogen, and ovalbumin. Biochemistry 1991; 30:1723-1730.
18. Lawrence DA, Olson ST, Palaniappan S, Ginsburg D. Serpin reactive center loop mobility is required for inhibitor function but not for enzyme recognition. J Biol Chem 1994; 269:27657-27662.
19. Wei A, Rubin H, Cooperman BS, Christianson DW. Crystal structure of an uncleaved serpin reveals the conformation of an inhibitory reactive loop. Nat Struct Biol 1994; 1:251-258.
20. Wei A, Rubin H, Cooperman BS, Schechter N, Christianson DW. Crystallization, activity assay and preliminary X-ray diffraction analysis of the uncleaved form of the serpin antichymotrypsin. J Mol Biol 1992; 226: 273-276.
21. Fitzpatrick PA, Wong,DT, Barr PJ, Pemberton PA. Functional implications of the modeled structure of maspin. Protein Eng 1996; 9:585-589.
22. Carrell RW, Evans DL, Stein PE. Mobile reactive centre of serpins and the control of thrombosis. Nature 1991; 353:576-578.
23. Hopkins PC, Stone SR. The contribution of the conserved hinge region residues of alpha1-antitrypsin to its reaction with elastase. Biochemistry 1995; 34:15872-15879.
24. Hopkins PC, Whisstock J. Function of maspin. Science 1994; 265:1893-1894.

25. Liu T, Pemberton PA, Robertson AD. Three-state unfolding and self-association of maspin, a tumor-suppressing serpin. J Biol Chem 1999; 274:29628-29632.
26. Stein PE, Carrell RW. What do dysfunctional serpins tell us about molecular mobility and disease? Nat Struct Biol 1995; 2:96-113.
27. Carrell RW, Stein PE, Fermi G, Wardell MR. Biological implications of a 3 A structure of dimeric antithrombin. Structure 1994; 2:257-270.
28. Evans DL, Marshall CJ, Christey PB, Carrell, RW. Heparin binding site, conformational change, and activation of antithrombin. Biochemistry 1995; 34:3478.
29. Hopkins PC, Pike RN, Stone SR. Evolution of serpin specificity: Cooperative interactions in the reactive-site loop sequence of antithrombin specifically restrict the inhibition of activated protein C. J Mol Evol 2000; 51:507-515.
30. Hopkins PC, Chang WS, Wardell MR, Stone SR. Inhibitory mechanism of serpins. Mobility of the C-terminal region of the reactive-site loop. J Biol Chem 1997; 272:3905-3909.
31. Pemberton PA, Tipton AR, Pavloff N, Smith J, Erickson JR, Mouchabeck ZM et al. Maspin is an intracellular serpin that partitions into secretory vesicles and is present at the cell surface. J Histochem Cytochem 1997; 45:1697-1706.
32. McGowen R, Biliran H, Sager R, Sheng S. The surface of prostate carcinoma DU145 cells mediates the inhibition of urokinase-type plasminogen activator by maspin. Cancer Res 2000; 60:4771-4778.
33. Grailhe P, Nieuwenhuizen W, Angles-Cano E. Study of tissue-type plasminogen activator binding sites on fibrin using distinct fragments of fibrinogen. Eur J Biochem 1994; 219:961-967.
34. Downing AK, Driscoll PC, Harvey TS, Dudgeon TJ, Smith BO, Baron M et al. Solution structure of the fibrin binding finger domain of tissue-type plasminogen activator determined by 1H nuclear magnetic resonance. J Mol Biol 1992; 225:821-833.
35. Moser TL, Enghild JJ, Pizzo SV, Stack MS. The extracellular matrix proteins laminin and fibronectin contain binding domains for human plasminogen and tissue plasminogen activator. J Biol Chem 1993; 268:18917-18923.
36. Baker MS, Liang XM, Doe WF. Occupancy of the cancer cell urokinase receptor (uPAR): Effects of acid elution and exogenous uPA on cell surface urokinase (uPA). Biochim Biophys Acta 1992; 1117:143-152.
37. Burgle M, Koppitz M, Riemer C, Kessler H, Konig B, Weidle UH et al. Inhibition of the interaction of urokinase-type plasminogen activator (uPA) with its receptor (uPAR) by synthetic peptides. Biol Chem 1997; 378:231-237.
38. Bonavaud S, Charriere-Bertrand C, Rey C, Leibovitch MP, Pedersen N, Frisdal E et al. Evidence of a non-conventional role for the urokinase tripartite complex (uPAR/uPA/PAI-1) in myogenic cell fusion. J Cell Sci 1997; 110:1083-1089.
39. Blasi F. uPA, uPAR, PAI-1: Key intersection of proteolytic, adhesive and chemotactic highways? Immunol Today 1997; 18:415-417.
40. Look MP. Pooled analysis of uPA and PAI-1 for prognosis in primary breast cancer patients. EORTC Receptor and Biomarker Study Group. Int J Biol Markers 2000; 15:70-72.
41. Rabbani SA, Xing RH. Role of urokinase (uPA) and its receptor (uPAR) in invasion and metastasis of hormone-dependent malignancies. Int J Oncol 1998; 12:911-920.
42. Grondahl-Hansen J, Christensen IJ, Rosenquist C, Brunner N, Mouridsen HT, Dano K et al. High levels of urokinase-type plasminogen activator and its inhibitor PAI-1 in cytosolic extracts of breast carcinomas are associated with poor prognosis. Cancer Res 1993; 53:2513-2521.
43. Stetler-Stevenson WG, Aznavoorian S, Liotta LA. Tumor cell interactions with the extracellular matrix during invasion and metastasis. Annu Rev Cell Biol 1993; 9:541-573.
44. Tarui T, Mazar AP, Cines DB, Takada Y. Urokinase receptor (uPAR/CD87) is a ligand for integrins and mediates cell-cell interaction. J Biol Chem 2000; 276:3983-3990.
45. Preissner KT, Kanse SM, May AE. Urokinase receptor: A molecular organizer in cellular communication. Curr Opin Cell Biol 2000; 12:621-628.

46. Chavakis T, Kanse SM, Lupu F, Hammes HP, Muller-Esterl W, Pixley RA et al. Different mechanisms define the antiadhesive function of high molecular weight kininogen in integrin- and urokinase receptor-dependent interactions. Blood 2000; 96:514-522.
47. Simon DI, Wei-Y, Zhang L, Rao NK, Xu H, Chen Z et al. Identification of a urokinase receptor-integrin interaction site. Promiscuous regulator of integrin function. J Biol Chem 2000; 275:10228-10234.
48. Pedersen AN, Christensen IJ, Stephens RW, Briand P, Mouridsen HT, Dano K et al. The complex between urokinase and its type-1 inhibitor in primary breast cancer: Relation to survival. Cancer Res 2000; 60:6927-6934.
49. Foekens JA, Peters HA, Look MP, Portengen H, Schmitt M, Kramer MD et al. The urokinase system of plasminogen activation and prognosis in 2780 breast cancer patients. Cancer Res 2000; 60:636-643.
50. Fujii T, Obara T, Tanno S, Ura H, Kohgo Y. Urokinase-type plasminogen activator and plasminogen activator inhibitor-1 as a prognostic factor in human colorectal carcinomas. Hepatogastroenterology 1999; 46:2299-2308.
51. Bajou K, Noel A, Gerard RD, Masson V, Brunner N, Holst-Hansen C et al. Absence of host plasminogen activator inhibitor 1 prevents cancer invasion and vascularization. Nat Med 1998; 4:923-928.
52. Germer M, Kanse SM, Kirkegaard T, Kjoller L, Felding-Habermann B, Goodman S et al. Kinetic analysis of integrin-dependent cell adhesion on vitronectin—The inhibitory potential of plasminogen activator inhibitor-1 and RGD peptides. Eur J Biochem 1998; 253:669-674.
53. Ehrlich HJ, Gebbink RK, Keijer J, Linders M, Preissner KT, Pannekoek H. Alteration of serpin specificity by a protein cofactor. Vitronectin endows plasminogen activator inhibitor 1 with thrombin inhibitory properties. J Biol Chem 1990; 265:13029-13035.
54. Laug WE, Cao X, Yu YB, Shimada H, Kruithof EK. Inhibition of invasion of HT1080 sarcoma cells expressing recombinant plasminogen activator inhibitor 2. Cancer Res 1993; 53:6051-6057.
55. Ossowski L, Reich E. Antibodies to plasminogen activator inhibit human tumor metastasis. Cell 1983; 35:611-619.
56. Zhang M, Shi Y, Magit D, Furth PA, Sager R. Reduced mammary tumor progression in WAP-TAg/WAP-maspin bitransgenic mice. Oncogene 2000; 19:6053-6058.
57. Shao ZM, Nguyen M, Alpaugh ML, O'Connell JT, Barsky SH. The human myoepithelial cell exerts antiproliferative effects on breast carcinoma cells characterized by p21WAF1/CIP1 induction, G2/M arrest, and apoptosis. Exp Cell Res 1998; 241:394-403.
58. Sager R, Sheng S, Anisowicz A, Sotiropoulou G, Zou Z, Stenman G et al. RNA genetics of breast cancer: Maspin as paradigm. Cold Spring Harb Symp Quant Biol 1994; 59:537-546.

Chapter 6

Genetic and Epigenetic Regulation of *Maspin* Gene Expression in Normal and Tumor Tissue

Frederick E. Domann and Bernard W. Futscher

Introduction

Maspin, a tumor suppressor gene, encodes a protein that has been shown to restrict breast cancer cell motility, invasion, and metastasis. Expression of the maspin gene is commonly silenced during breast cancer progression, and this loss of expression has been shown to occur at the level of transcription. Silencing of maspin expression in human breast cancer is frequently associated with aberrant cytosine methylation, histone hypoacetylation, and chromatin condensation in the maspin promoter region. The epigenetic changes associated with maspin gene silencing in human breast cancer cells suggest that this gene may be a viable target for pharmacological reactivation. Indeed, maspin gene expression can be reactivated in human breast cancer cells by 5-aza-2'deoxycytidine and trichostatin A, two drugs that inhibit DNA methyltransferase and histone deacetylase, respectively. A deeper understanding of the mechanisms underlying regulation of maspin gene expression in normal and cancer cells will be important in developing new targeted therapies against cancer.

Maspin is a Member of the Serpin Family of Serine Protease Inhibitors

Maspin, also called proteinase inhibitor 5 (PI5) (OMIM # 154790), is a member of the serpin family of proteins.[1] There are an estimated 30 to 40 serpins, many occurring in clusters on different chromosomes.[2-4] The maspin gene is located in a serpin cluster located in a 500-kb region of 18q21. There are 6 serpins in this cluster that are from the ovalbumin family of serpins (Ov-serpins). In addition to maspin, plasminogen activator inhibitor type 2 (PAI2), squamous cell carcinoma antigen-1 (SCCA1) and -2 (SCCA2), cytoplasmic antiproteinase 2 (PI8) and bomapin (PI10) are also located in this cluster. From centromere to telomere the order is PI5—SCCA1—SCCA2—PAI2—PI10—PI8.[5,6] These genes contain seven or eight exons, and have similar exon/intron structures, and it is thought that these genes arose from duplication events.[3] Interestingly, while there is overlap in the tissue specific expression of some of these serpins, this overlap is not absolute. For instance, maspin is expressed in breast but not bone marrow or heart, while bomapin is expressed in bone marrow but not breast or heart, and PI8 is widely expressed, including heart. This suggests that in addition to divergent evolution of the proteins in this cluster, the regulatory regions of these genes are also likely to be different. One such difference is that it appears that the maspin 5' promoter region is the only CG-rich promoter in this serpin cluster.

Maspin Expression in Normal and Cancerous Tissue

As suggested above, maspin gene expression appears to be restricted to epithelial cells. In addition to breast epithelium, maspin expression is also seen in prostate, skin, lung, and GI epithelial cells, but maspin expression is not found in a majority of tissues that includes heart, peripheral blood, brain, or fibroblasts derived from breast or skin[7] (and unpublished observations). Loss or decreased levels of maspin gene expression is a frequent event during breast cancer progression, and has also been described in prostate and oral squamous cell carcinomas.[8,9] Interestingly, no mutations in the maspin coding sequence nor gross deletions or rearrangements of the gene have been identified in human breast cancer cells that could account for the loss of maspin gene expression.[1] In human prostate cancer cell lines and tumor specimens, variations from the published maspin cDNA sequence[1] (Genbank Accession No. U04313) have been identified; however, these changes produce conservative amino acid changes,[10] and the functional implications of these changes are currently unknown. As such, it remains unclear if these changes are true mutations or simply allelic variation.[10] Thus, although it is clear that maspin gene silencing is a common event during the progression of some epithelial tumors, the mechanism(s) responsible for maspin gene silencing has remained elusive.

Transcriptional Regulatory Elements in the 5' Flanking Region of the Maspin Gene

Analysis of the maspin gene 5' flanking region reveals numerous possible cis-elements that may participate in its transcriptional regulation (see Fig. 1). These include, but are not limited to, p53, AP1, Ets, HIF1, AP2, and nuclear hormone receptors. Of these, AP1, Ets, and p53 are the best analyzed to date. Using maspin promoter reporter constructs transfected into breast cells, Zhang et al demonstrated that site directed mutation of the AP1 or ets sites most proximal to the transcription start site were capable of greatly reducing maspin promoter activity.[11] Later studies by this same group showed that the ets site also positively regulated maspin promoter activity in prostate cells.[8] This is further supported by the recent isolation and identification of a prostate specific ets from mouse that is capable of stimulating the human maspin promoter.[12]

More recently, p53 has been implicated as a transcriptional regulatory factor governing the expression of maspin.[13] Zou et al used an adenoviral expression system to over-express wild type p53 in two human prostate cancer cell lines, PC-3 (p53 null) and LNCaP(p53 wild type) and one breast cancer cell line, MCF-7 (p53 wild type). Adenovirus mediated p53 over-expression led to significant accumulation of maspin mRNA in both of the prostate cancer cell lines. In contrast, MCF-7 breast cancer cells

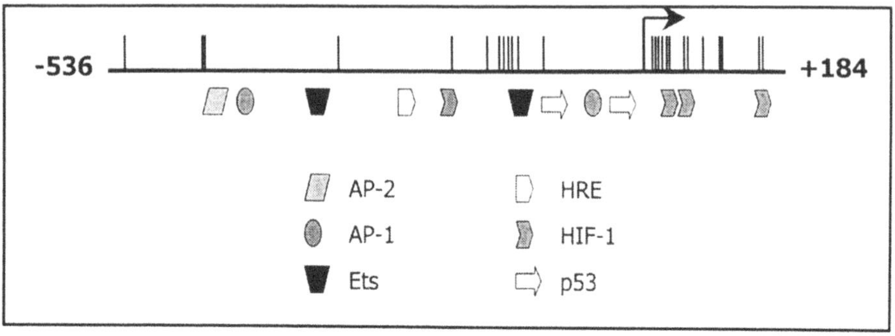

Fig. 1. A map of the maspin promoter showing landmarks and features discussed within the text. CpG dinucleotides are indicated as vertical tick marks, while the transcription start site is indicated with a bent arrow. Locations for known and putative cis-elements within the promoter region are also shown. These include AP-1, AP-2, Ets, HRE, p53, and HIF-1.

displayed only a minimally increased accumulation of maspin mRNA upon enforced p53 over-expression. Strong activation of the maspin promoter in reporter gene assays was also observed in both human prostate cancer cell lines. This activation of the maspin promoter was accompanied by increased binding to two putative p53 responsive cis-elements in the maspin promoter. The reasons for the difference between p53 mediated transactivation of maspin expression in the prostate and breast cancer cell lines are at present unclear, however they may reflect differences in the underlying mechanisms of transcriptional silencing between these two tumor types.

Overall, studies on the mechanisms underlying the loss of maspin expression have converged on its transcriptional regulation as the primary mode of gene silencing. While the ets response element appears to positively regulate the maspin promoter, ets family members are often overexpressed in breast cancer,[14] thereby limiting the likelihood that loss of ets activity is responsible for maspin gene silencing. Similarly, while the AP1 response element positively regulates the maspin promoter, increased AP1 activity is typically associated with tumor progression,[15,16] although it has been reported that AP1 activity may sometimes be decreased in breast cancer cells.[17] So, while decreased levels of AP1 may lead to a down-regulation of maspin gene expression, it seems unlikely that it would be responsible for complete silencing. Since p53 has been shown to activate the maspin promoter in human prostate cancer cells, and p53 is affected in about 60% of all human cancers,[13] it is possible that loss of wild type p53 function may contribute to loss of maspin expression. However, breast cancer cell lines that have wild type p53, such as MCF-7, still fail to express maspin, and even enforced p53 over-expression in MCF-7 leads to only a modest accumulation of maspin mRNA.[13] As such, it appears that while changes in the status of important transcription factors undoubtedly participate in the silencing of maspin gene expression, it seems unlikely to be a unifying mechanism. In the proceeding sections we will briefly describe our hypothesis that epigenetic mechanisms play an important and significant role in the silencing of maspin expression in human breast cancer.

Table 1. Loss of maspin mRNA expression is frequently associated with maspin gene promoter methylation in a diverse array of human normal breast and breast cancer cell lines.

Cell line	Maspin Expression	Promoter Methylation
HMEC	++++	--
MCF-10A	++++	--
UACC 245	+++	--
ZR-75.1	+	-
MCF-7	--	+++
BT549	--	++
Hs578T	--	++
UACC 1179	--	+++
UACC 893	--	++
UACC 3133	--	+
MDA-MB-231	--	++
MDA-MB-435	--	+++
UACC 2087	--	-
PBL	--[a]	+++

[a] Pemberton et al, 1997

Review of Current Research

Mechanisms of Silencing Maspin Gene Expression in Human Cancer

Analysis of the maspin promoter region from –176 to +158 reveals that it has a GC content of >50% and a frequency of the dinucleotide 5'-CpG-3' three times greater than the bulk of the genome, thereby fitting the mathematical definition of a CpG island.[18] Pioneering work by Baylin and colleagues has clearly demonstrated that aberrant methylation of cytosines in CpG dinucleotides in the promoter of tumor suppressor genes is an important mechanism of gene silencing in human cancer.[19] Since maspin functions as a class two tumor suppressor in human breast cancer, and is frequently targeted for gene silencing during cancer progression, we hypothesized that the maspin gene might be a target of methylation-associated silencing.[7]

To initiate the test of this hypothesis, we compared maspin mRNA expression in a panel of human breast cancer cell lines to normal human mammary epithelial cells. Our original panel consisted of a number of cell lines derived at the Arizona Cancer Center (UACC cell lines) as well as a number of more commonly used breast cancer cell lines including MCF-7, ZR-75-1, BT549, and Hs578T.[7] To determine relative levels of maspin mRNA expression we have used both northern blotting and semi-quantitative RT-PCR. Our results from these northern blot analyses, a representation of which are shown in Figure 2, have confirmed and extended previous studies that showed maspin mRNA expression is absent or vastly decreased in the majority of advanced breast cancers. Since then we have expanded our panel of analyzed cell lines to include MDA-MB-231and MDA-MB-435 human breast cancer cell lines, which

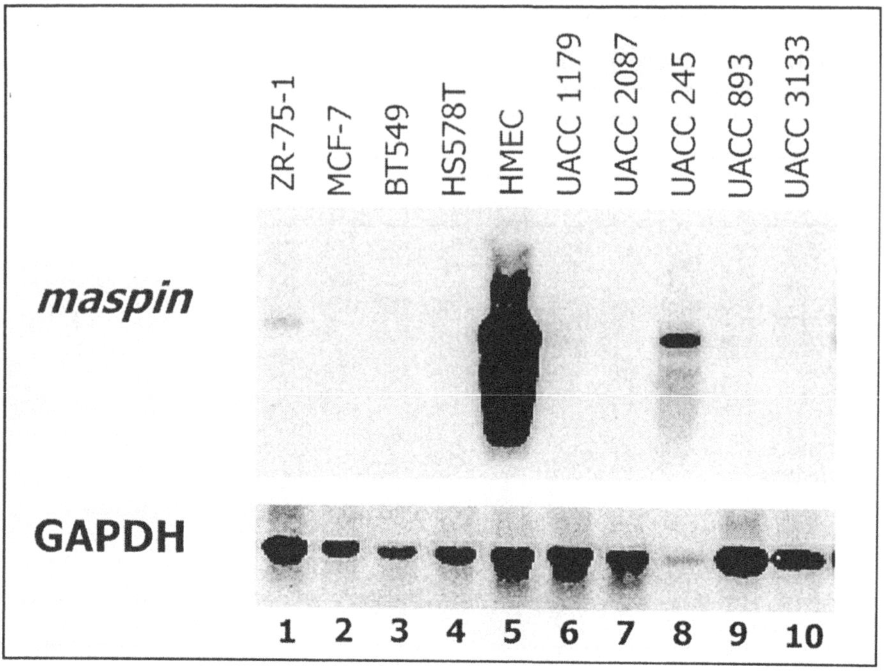

Fig. 2. Northern blot analysis of maspin expression in normal human mammary epithelial cells and breast cancer cells. Ten micrograms of total cellular RNA were electrophoresed in each lane. Lane 1, ZR-75-1; lane 2, MCF-7; lane 3, BT549; lane 4, HS578T; lane 5, HMEC; lane 6, UACC1179; lane 7, UACC2087; lane 8, UACC245; lane 9, UACC893; lane 10, UACC3133. Top panel, maspin; bottom panel, GAPDH. (Adapted from Domann FE et al. Epigenetic silencing od maspin gene expression in human breast cancers. Internat J Can 2000; 85:805-810. Copyright © 2000, Wiley-Liss, Inc., a subsidiary of John Wiley and Sons, Inc.)

are maspin-negative, and the spontaneously immortalized but non-tumorigenic human MCF-10A cell line, which is maspin-positive. Results from these studies are summarized in Table 1.

Loss of Maspin Expression in Human Breast Cancer Cells is Associated with Aberrant Hypermethylation of the Maspin Promoter

Bisulfite genomic sequencing was used to analyze and quantify the degree of cytosine methylation at CpG dinucleotides in the maspin promoter. This method chemically converts unmethylated cytosine to uracil (which PCR amplifies as thymidine), but does not convert 5-methylcytosine.[20] This allows for the discrimination between methylated and non-methylated cytosines in genomic DNA. After bisulfite conversion the region of interest, in this case a region surrounding the maspin transcription start site, is amplified by PCR, and ligated into a TA vector for transformation of competent bacteria. For our analyses we sequenced ten individual randomly chosen

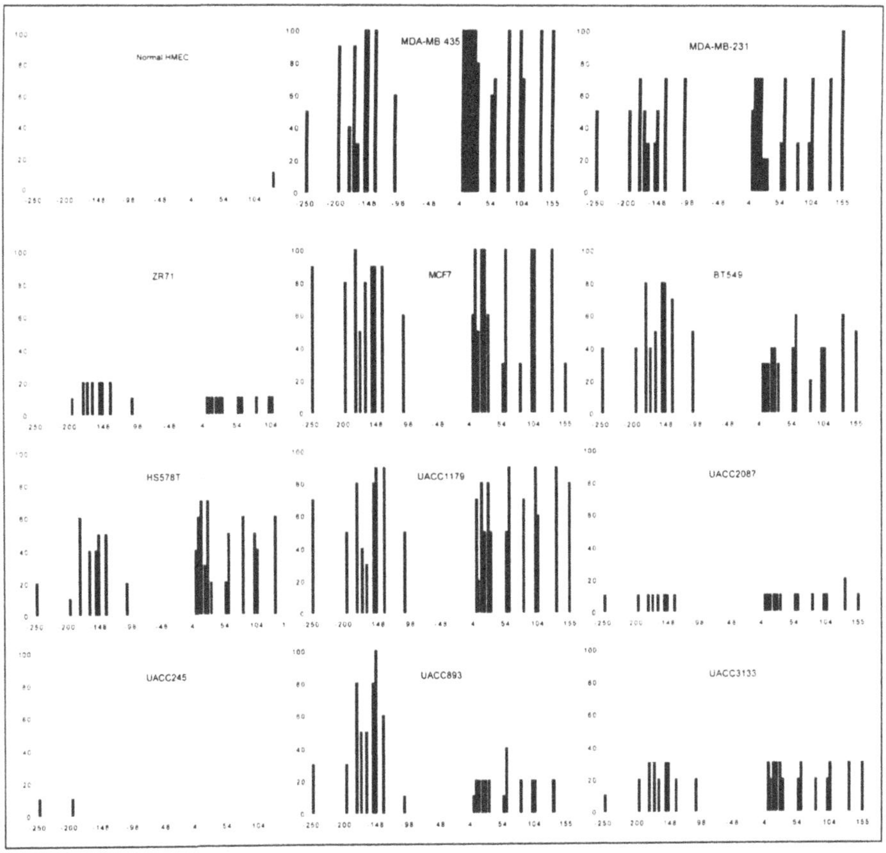

Fig. 3. The maspin promoter is frequently methylated in human breast cancer cells with low or undetectable maspin mRNA expression. Ten cloned PCR products were sequenced to determine the percent methylation of the 21 CpG sites in the region analyzed. Cytosine methylation frequency histograms are shown for normal HMECs and each of the eleven human breast cancer cell lines examined. The x-axis is the nucleotide position relative to the transcription start and the y-axis is the percent cytosine methylation at any given CpG. (Adapted from Domann FE et al. Epigenetic silencing od maspin gene expression in human breast cancers. Internat J Can 2000; 85:805-810. Copyright © 2000, Wiley-Liss, Inc., a subsidiary of John Wiley and Sons, Inc.)

clones to determine the cytosine methylation frequency at 21 CpG sites around the maspin transcriptional start site. Results of these experiments are shown in Figure 3 and summarized with the expression data in Table 1. It is clear from these findings that loss of maspin expression is commonly associated with aberrant cytosine methylation in the maspin promoter region. Among nine human breast cancer cell lines that lacked detectable maspin mRNA expression by northern blot analysis, eight of them showed a significant degree of cytosine methylation within their 5' regulatory region. In contrast, in normal human mammary epithelial cells (HMEC) and three breast cancer cell lines in which maspin mRNA was detectable, the maspin gene 5' regulatory region remained unmethylated.[7] Interestingly, a region further upstream of the transcription start site, from nucleotide – 950 to – 450, was methylated to a similar

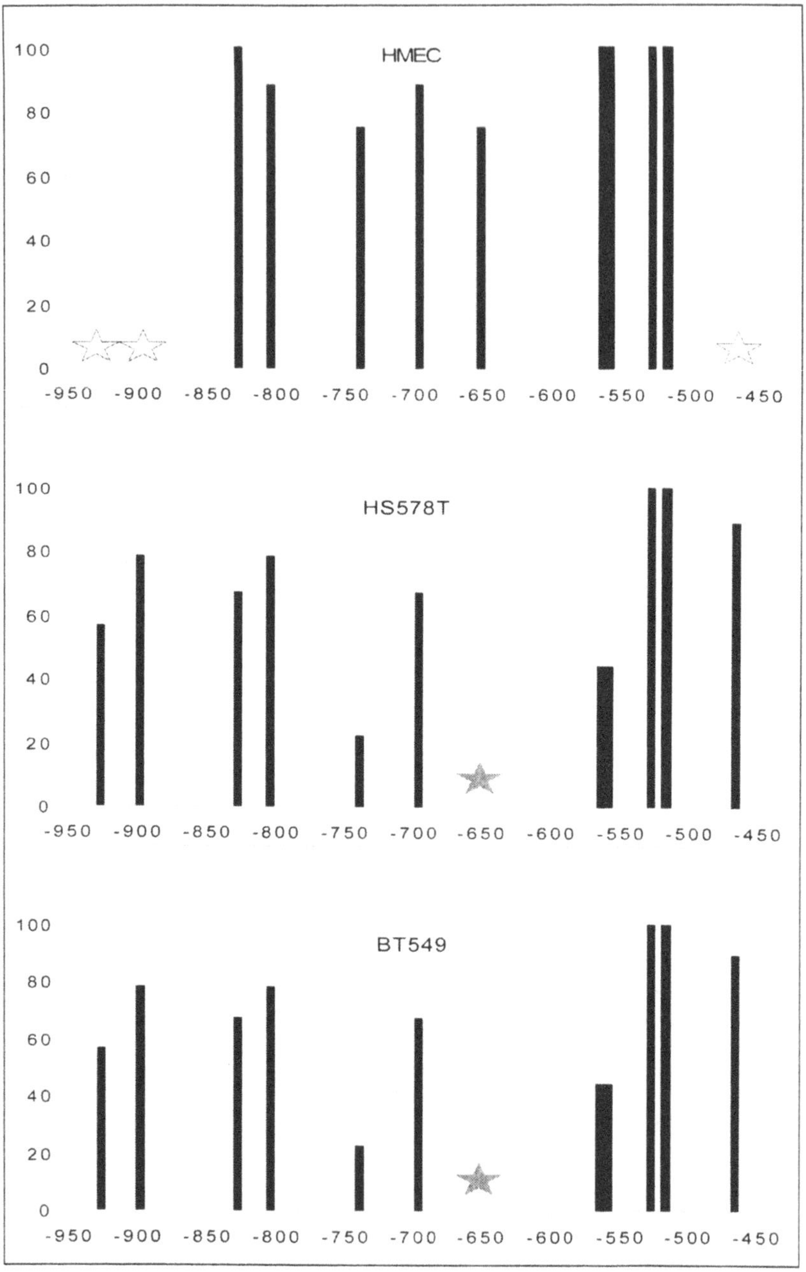

Fig. 4. Summary of 5-methylcytosine levels obtained by sodium bisulfite genomic sequencing in the 550 bp upstream portion of the maspin promoter. Ten cloned PCR products were sequenced to determine the percent methylation of the 13 CpG sites in the region analyzed. Cytosine methylation frequency histograms are shown for the normal maspin-positive HMECs and the maspin-negative breast cancer cell lines, HS578T and BT549. The x-axis is the nucleotide position relative to the transcription start, and the y-axis is percent cytosine methylation at each CpG site. Open star, CpG sites where methylation status was not determined; shaded star, CG ‡ TA transitions detected.

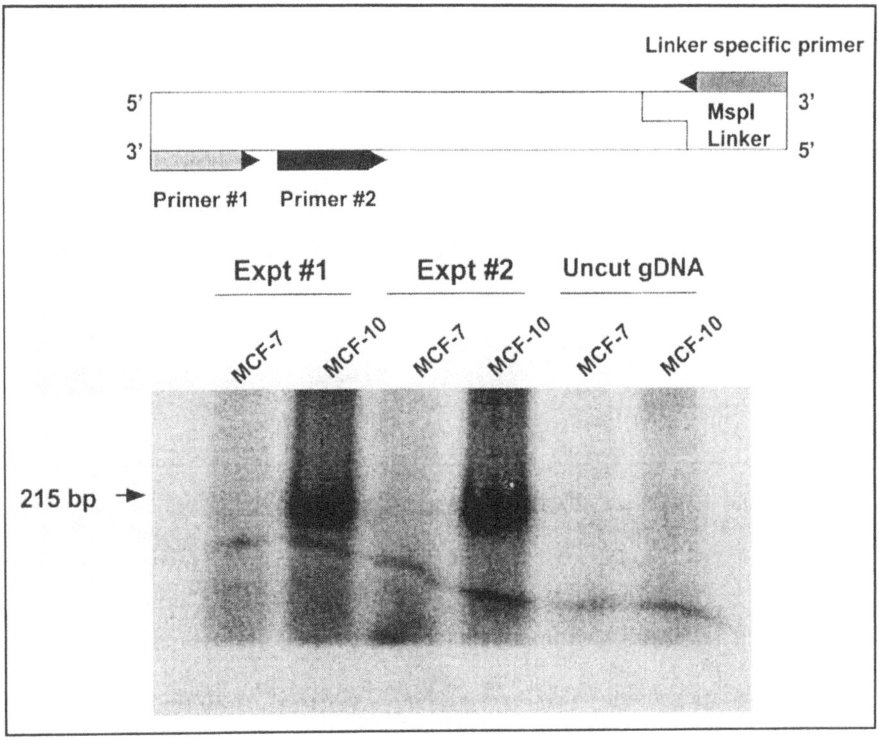

Fig. 5. The methylated maspin promoter adopts a closed chromatin structure. Nuclei from MCF10A maspin positive breast cells and MCF7 maspin negative breast cancer cells were exposed to 75 units of Msp I restriction endonuclease to assess in vivo accessibility of the maspin promoter. The genomic DNA was isolated, ligated to MspI-specific linkers, and PCR-amplified with a maspin promoter-specific primer and a linker-specific primer for 20 cycles. A second, ^{32}P-labeled maspin promoter-specific primer internal to the first primer was added to the PCR reaction and an additional 3 cycles of PCR were performed. The PCR products were then separated on a native on a 6% polyacrylamide gel. The presence of a 215 bp maspin promoter-specific PCR product indicates that the MspI site was cleaved by MspI, and therefore the maspin promoter has an accessible configuration in the maspin positive MCF10A cells. The absence of a 215 bp band in the maspin negative MCF7 cells indicates that the maspin promoter has an inaccessible chromatin configuration in these cells. The results from 2 independent experiments are shown.

extent in both maspin expressing normal human mammary epithelial cells as well as non-expressing breast carcinoma cell lines Hs578T and BT549 (Fig. 4). This finding suggests that aberrant methylation localized to the proximal 5' regulatory region of the maspin may be of major importance in silencing of maspin expression.

The Maspin Promoter Resides in a Heterochromatic State in Cells That Do Not Express Maspin

Several proteins have been identified that bind to methylated CpGs and these proteins have been reported to recruit histone deacetylases to heavily methylated DNA.[21] Recruitment of histone deacetylases to a particular region of DNA would be expected to result in the formation of heterochromatin, a chromatin state that is in-

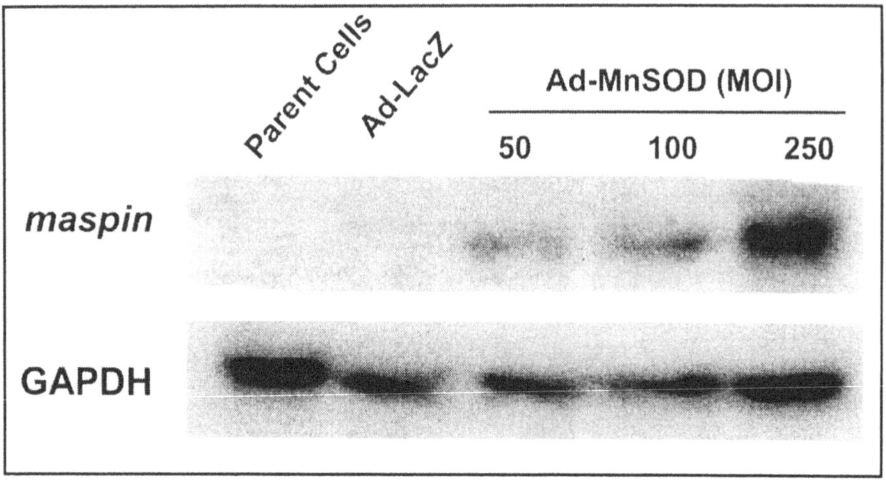

Fig. 6. Enforced expression of manganese superoxide dismutase (MnSOD) causes accumulation of maspin mRNA in MDA-MB-231 human breast cancer cells in a dose dependent manner as determined by northern blot analysis. Cells were incubated with increasing numbers of adenovirus particles per cell (multiplicities of infection, MOI) in serum free medium for 18 hours. The medium was replaced with serum containing medium and the cells were incubated for an additional 24 h before RNA was isolated. AdLacZ is an adenovirus encoding lacZ as a control for the effects of adenoviral infection.

hibitory to transcription. To determine whether the CpG methylated region around the maspin transcription start site in non-expressing cells was maintained in a transcription factor inaccessible state, we performed chromatin accessibility assays on isolated nuclei from maspin-positive MCF10A cells and from maspin-negative MCF-7 breast cancer cells.[7] The isolated nuclei from these cells were exposed to the restriction endonuclease MspI, which is capable of endonucleolytic cleavage at one site within the region of the human maspin promoter analyzed (see Fig. 1). Following this in vivo endonuclease digestion, the genomic DNA was extracted from the isolated nuclei, and chromatin accessibility of the maspin promoter was assayed using a modification of the linker-mediated PCR protocol originally described by Pfeiffer and Riggs.[22] Incubation of the MCF10A nuclei with 75 units of MspI resulted in cleavage at the maspin promoter MspI site as evidenced by the presence of a PCR product at 215 bp (Fig. 5). In contrast, when the nuclei from the maspin-negative MCF7 were exposed to 75 units of MspI no cleavage of the maspin promoter could be detected, as evidenced by the lack of the maspin promoter specific PCR product at 215 bp. The results of these experiments indicate that the chromatin around the maspin promoter is maintained in an open structure in normal maspin-positive MCF10A cells, whereas this region has adopted an inaccessible structure in the region of the maspin promoter that is aberrant methylated in the maspin-negative MCF-7 breast cancer cells.

Maspin mRNA Expression Can Be Reactivated in Maspin-Negative Cells

The frequent loss of maspin expression in breast cancer and its resultant effects on the breast cancer phenotype make maspin replacement a potentially attractive therapeutic strategy. Approaches could include treatment with recombinant maspin protein or small molecule maspin-mimetics, gene therapy protocols designed to reintroduce an active maspin gene into the tumor, or reactivation of the silenced maspin gene. To date, it has been reported that gamma linolenic acid is capable of inducing maspin expression and maspin protein levels in MCF7 cells, but not MDA-MB-231 cells.[23] In addition, Li et al first reported the ability of manganese superoxide dismutase (MnSOD) over-expression by stable transfection to induce maspin gene expression in the human breast cancer cell line MCF7 [24]. Further studies using adenovirus mediated gene transfer of MnSOD have extended these findings to MDA-MB-231 and other human breast cancer cell lines (Fig. 6). Together these results suggest that molecular targets of essential fatty acids and/or antioxidants may be regulators of maspin gene expression. These inducers of maspin expression are not known to influence DNA methylation and therefore are unlikely to induce maspin gene expression by demethylation of the maspin promoter. This data suggests that these compounds either greatly stimulate the very small percentage of cells that have a relatively unmethylated maspin promoter or that aberrant methylation is not sufficient to completely repress maspin gene expression.

Pharmacologic approaches have also been used to determine whether aberrant methylation of the maspin promoter participates in the inappropriate silencing of maspin. Maspin-negative, aberrantly methylated MCF7 and MD-MBA-435 breast cancer cells were exposed to varying concentrations of the C5-DNA methyltransferase inhibitor, 5-aza-2'-deoxycytidine (5-aza-dC). Both cell lines were capable of inducing maspin expression following exposure to 5-aza-dC, suggesting that demethylation of the maspin promoter can allow for the re-expression of maspin. A dose response curve of maspin induction versus dose of 5-aza-dC for MCF7 is shown in Figure 7. The results presented are the mean fold-induction of maspin from 3 independent experiments. The diminution of maspin induction at higher concentrations of 5-aza-dC appears to be a result of drug-induced cytotoxicity. Further support, albeit indirect, for a role of cytosine methylation in the regulation of maspin gene expression comes from studies on malignant oral keratinocytes.[25] Transfection of malignant oral keratinocytes with the ornithine decarboxylase antizyme resulted in decreased levels of genomic methylation that more closely approximated levels seen in normal oral keratinocytes. This decreased level of overall methylation was associated with a significant decrease in tumor size in mouse xenografts, as well as an increase in maspin gene expression. Together, these experiments suggest that aberrant methylation of the maspin promoter participates in the silencing of maspin gene expression.

Cytosine methylation of promoters may act to silence gene expression directly by sterically hindering the interaction between a cis-response element and its cognate transcription factor, or indirectly where methylation of the promoter mediates a structural change that results in a chromatin structure that occludes transcription factors that are present from binding their cognate site. Recently, Bird's group and Wolffe's group have reported powerful studies that show the methylated DNA binding protein MeCP2 can recruit histone deacetylase (HDAC) to the region and initiate

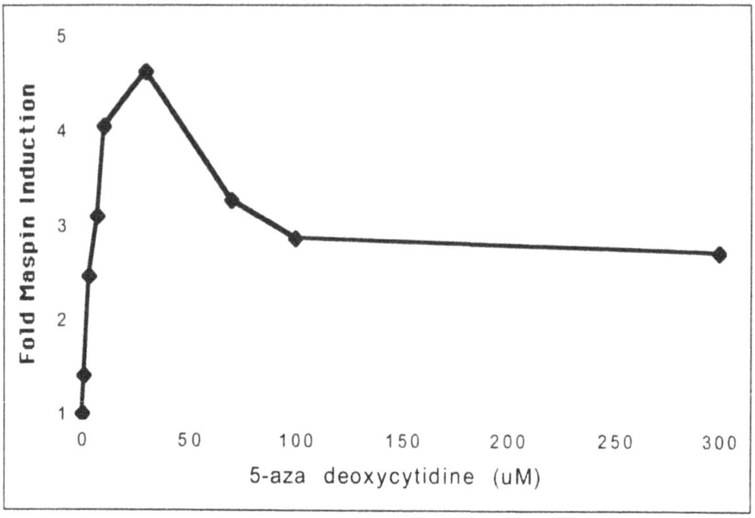

Fig. 7. Dose dependent re-activation of maspin expression in MCF-7 cells treated with 5-aza-2'-deoxycytidine. Aberrantly methylated, maspin-negative MCF7 cells were exposed to increasing concentrations of 5-aza-2'-deoxycytidine (5-aza-dC) for 4 days. Total RNA was isolated and maspin expression was analyzed by quantitative RT-PCR analysis. Data shown are the mean of three independent experiments. 5-aza-dC induced a marked dose dependent increase that peaked at about 30 μM. Decreases at higher doses were likely due to cytotoxicity of drug.

heterochromatin formation and transcriptional silencing.[26] Based on these results and the studies on maspin transcriptional regulation to date, we favor a model for maspin gene silencing in breast cancer where methylation of the maspin promoter results in the recruitment of factors that affect chromatin structure, such as post-translational histone modification. If this model is correct, then therapeutic agents that affect either cytosine methylation or histone acetylation would be expected to be effective at reactivating maspin expression in breast cancer cells. When used in judicious combinations, these pharmacologic strategies designed to 1) overcome the repressive effects of aberrant DNA methylation pathways, and 2) stimulate positive regulators of maspin gene expression, may provide opportunities to maximize gene induction and minimize drug exposure.

Future Directions

Does DNA Methylation Affect Normal Tissue Restricted Expression of Maspin?

In our initial analysis of maspin expression and promoter methylation state in human breast cancer cell lines we also included normal human peripheral blood lymphocytes. Surprisingly, while the maspin promoter was completely unmethylated in normal maspin-positive breast epithelial cells, the maspin promoter was completely methylated in normal, but maspin-negative peripheral blood mononuclear cells (Fig. 8). Thus, it appears the maspin promoter is not a typical CpG island. Specifically, while the maspin promoter fits the mathematical criteria of a CpG island[18] as well as

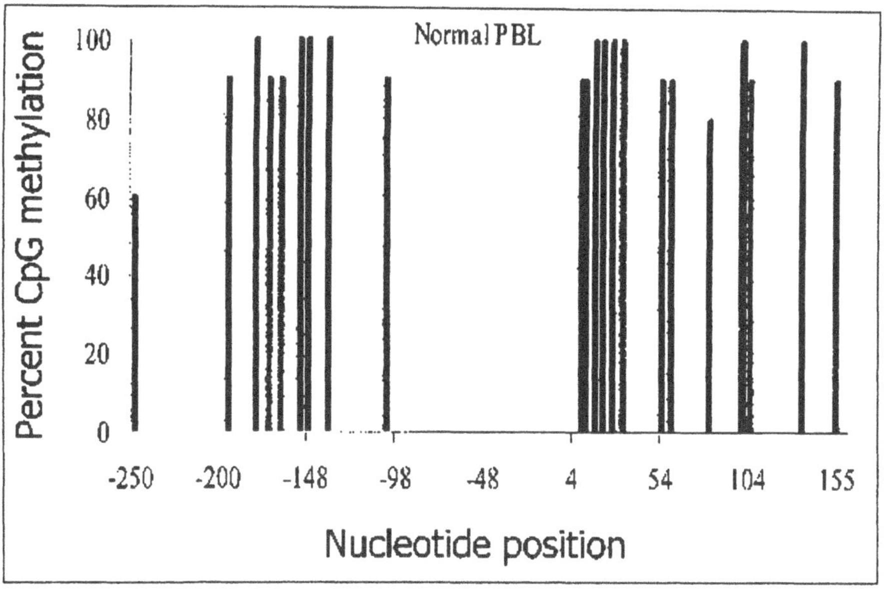

Fig. 8. The maspin promoter is methylated in normal human peripheral blood lymphocytes that display undetectable maspin mRNA expression. Similar results have been obtained from normal human heart, kidney, as well as various tissue-derived fibroblasts in which maspin expression is negative. Ten cloned PCR products were sequenced to determine the percent methylation of the 21 CpG sites in the region analyzed. Cytosine methylation frequency histograms are shown. The x-axis is the nucleotide position relative to the transcription start and the y-axis is the percent cytosine methylation at any given CpG. Adapted from Domann FE et al. Epigenetic silencing od maspin gene expression in human breast cancers. Internat J Can 2000; 85:805-810. Copyright © 2000, Wiley-Liss, Inc., a subsidiary of John Wiley and Sons, Inc.

one functional characteristic, namely that it overlaps the transcriptional start site, the maspin promoter does not meet another functional characteristic – that a CpG island is unmethylated in all normal tissue regardless of gene expression status.[27,28]

The surprising results from the peripheral blood lymphocytes suggested that cytosine methylation of the maspin promoter may be a normal epigenetic mechanism involved in the tissue-restricted expression of the maspin gene, and that this process might be recapitulated in breast cancer cells as they progress to a more aggressive, less differentiated state. To initiate a test of the hypothesis that DNA methylation is involved in the control of the normal tissue-restricted expression of maspin, we have begun to analyze the maspin expression status and promoter methylation pattern in a variety of normal human tissues. Our preliminary findings suggest that there is an absolute and inverse correlation between maspin mRNA expression and the cytosine methylation status of the maspin promoter in normal human cells. For example, in normal human epithelial cells from breast, prostate, skin, airway, and oral cavity, maspin mRNA is abundantly expressed and the maspin 5' regulatory region is completely unmethylated (unpublished observations). In contrast, in normal human peripheral blood, kidney, heart, as well as various tissue-derived fibroblasts maspin mRNA is undetectable and the maspin promoter is universally methylated (Fig. 8 and unpublished observations). Thus, tissue specific expression of the maspin gene closely mirrors the methylation status of the promoter within a given cell and tissue type. These findings suggest that tissue-restricted expression of maspin in adult tissues may be at least in part the result of maspin promoter methylation. This differential methylation of the maspin promoter in normal tissues might arise during normal development and may help to establish expression patterns of the maspin gene in all normal adult tissues. We speculate that the maspin locus may undergo the same types of epigenetic changes in normal cells during epithelial to mesenchymal transitions as it does in transformed cells during tumor progression. Moreover, this may be a common process during differentiation leading to tissue specific gene expression patterns. Thus, one of our future goals is to characterize the changing methylation status of the maspin promoter during development and differentiation.

Higher-Order Chromatin Structure in the Maspin 5' Regulatory Region

We have previously demonstrated that the maspin promoter adopts a closed chromatin structure in non-expressing and heavily methylated MCF-7 breast cancer cells compared to normal human mammary epithelial cells which are unmethylated, have an open chromatin structure and express high levels of maspin.[7] The connection between cytosine methylation, methylcytosine-binding proteins and histone deacetylases as means to affect chromatin structure has prompted an analysis of the histone acetylation state of the maspin promoter. To date, preliminary results obtained from chromatin immunoprecipitations using antibodies specific for acetylated histones H3 and H4 coupled to a PCR analysis of the maspin promoter region indicate that both of these histones are acetylated in unmethylated maspin-positive normal human breast epithelial cells, whereas both of these histones are deacetylated in the aberrantly methylated, maspin-negative breast cancer cells. Moreover, our preliminary findings suggest that the histone deacetylase inhibitor trichostatin A can serve as another pharmacological strategy to reactivate maspin transcription in human breast cancer cells. Figure 9 shows a schematic illustration of our interpretation of the maspin promoter in normal human breast epithelial cells compared to human breast cancer cells. In normal breast cells the maspin promoter is unmethylated, associated with acetylated histones, and in an open chromatin configuration that is accessible for interaction with transacting factors such as AP-1 and Ets. During malignant progression, the maspin promoter undergoes epigenetic changes including increased cytosine methylation, decreased histone acetylation, and chromatin condensation that render the promoter inaccessible for transcription factor binding and transactivation. The sum total of these epigenetic effects leads to the transcriptional silencing of the maspin gene in human breast cancers. The high frequency with which these changes are identified in human breast cancer cells suggests that this is a common mechanism for transcriptional silencing of the gene, and further suggests therapeutic strategies that may facilitate re-expression of the gene with the aim of slowing cancer progression.

Conclusion

Loss of maspin expression appears to be a common event during progression of not only human breast cancer, but other cancers as well. The maspin gene is down-regulated in human prostate cancer,[8] and its lack of expression in human oral cancers is associated with a poor prognosis.[9] In addition, other serpins besides maspin have

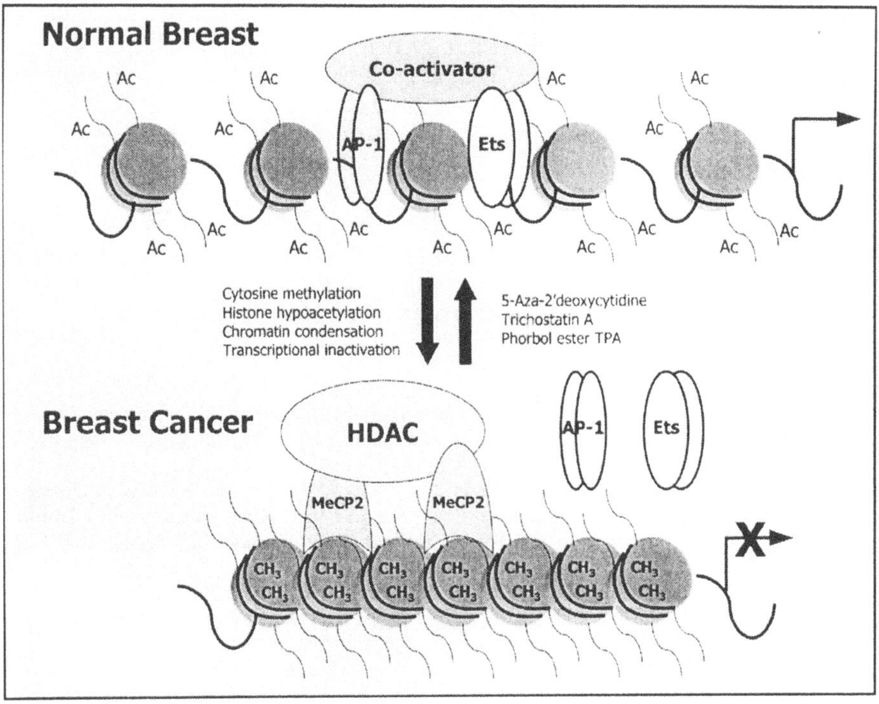

Fig. 9. Schematic illustration of the maspin promoter in normal human breast epithelial cells compared to human breast cancer cells. In normal breast cells the maspin promoter is unmethylated, associated with acetylated histones H3 and H4, and in an open chromatin configuration that is accessible for interaction with transacting factors such as AP-1 and Ets. During malignant progression, the maspin promoter undergoes epigenetic changes including increased cytosine methylation, decreased histone acetylation, and chromatin condensation that render the promoter inaccessible for transcription factor binding and transactivation. The sum total of these epigenetic effects leads to the transcriptional silencing of the maspin gene in human breast cancers, a common event in cancer progression.

been implicated as having a role in tumor progression. For example, headpin was found to be under-expressed in cancers of the head and neck[29] and pancpin was shown to be down-regulated in human pancreatic cancer.[30] Taken together, these observations suggest that at least certain members of the serpin family of proteins provide a critical function that inhibits or prevents cancer cells from progressing to a more aggressive invasive and metastatic phenotype. One such member of the serpin family, maspin, has taken center stage in the study of human breast cancer progression and provides an attractive molecular target for therapeutic intervention aimed at slowing disease progression.

In conclusion, maspin gene expression is frequently lost in human breast cancer; however, this loss of expression is not due to mutation, deletion, or rearrangement.[1] We propose that a major mechanism of maspin gene silencing in human breast cancer is not due to a difference in the abundance or activity of transcription factors per se, but rather to epigenetic events that occur around the maspin transcription start site.[7] These epigenetic events in transformed human mammary cells include aberrant DNA hypermethylation and histone hypoacetylation around the maspin transcription start site that cause the chromatin to adopt a closed structure that is inaccessible for efficient transcription (Fig. 9). These findings suggest that pharmacological strategies that affect these processes may be useful as therapies to reactivate maspin expression and slow the progression of human breast cancer.

Ackowledgements

This work was supported by Public Health Service grants CA73612 to FED and CA65662 to BWF. The authors would like to thank the various people from the Domann and Futscher labs that have contributed to this research. They are Hong Duan, Christin Hanagan, Nick Holtan, Marc Oshiro, and Ryan Wozniak.

References

1. Zou Z, Anisowicz A, Hendrix MJC, Thor A, Neveu N, Sheng S et al. *Maspin,* a serpin with tumor suppressing activity in human mammary epithelial cells. Science 1994; 263(5146):526-529.
2. Sun J, Stephens R, Mirza G, Kanai H, Ragoussis J, Bird PI. A serpin gene cluster on human chromosome 6p25 contains P16, P19 and ELANH2 which have a common structure almost identical to the 18q21 ovalbumin serpin genes. Cytogenet Cell Genet 1998; 82(3-4):273-277.
3. Scott FL et al. Human ovalbumin serpin evolution: Phylogenic analysis, gene organization, and identification of new PI8-related genes suggest that two interchromosomal and several intrachromosomal duplications generated the gene clusters at 18q21-q23 and 6p25. Genomics 1999; 62(3):490-499.
4. Chang WSW, Chang NT, Lin SC, Wu CW, Wu FYH. Tissue-specific cancer-related serpin gene cluster at human chromosome band 3q26. Genes Chromosomes Cancer 2000; 29(3):240-255.
5. Bartuski AJ, Kamachi Y, Schick C, Overhauser J, Silverman GA. Cytoplasmic antiproteinase 2 (PI8) and bomapin (PI10) map to the serpin cluster at 18q21.3. Genomics 1997; 43(3):321-328.
6. Silverman GA, Bartuski AJ, Cataltepe S, Gornstein ER, Kamachi Y, Schick C et al. SCCA1 and SCCA2 are proteinase inhibitors that map to the serpin cluster at 18q21.3. Tumour Biology 1998; 19(6):480-487.
7. Domann FE, Rice JC, Hendrix MJ, Futscher BW. Epigenetic silencing of *maspin* gene expression in human breast cancers. Internat J Can 2000; 85(6):805-810.
8. Zhang M, Magit D, Sager R. Expression of *maspin* in prostate cells is regulated by a positive ets element and a negative hormonal responsive element site recognized by androgen receptor. Proceedings of the National Academy of Sciences of the United States of America 1997; 94(11):5673-5678.
9. Xia W et al. High tumoral *maspin* expression is associated with improved survival of patients with oral squamous cell carcinoma. Oncogene 2000; 19(20):2398-2403.
10. Umekita Y, Hiipakka RA, Liao S. Rat and human *maspins*: Structures, metastatic suppressor activity and mutation in prostate cancer cells. Cancer Lett 1997; 113(1-2):87-93.
11. Zhang M, Maass N, Magit D, Sager R. Transactivation through ets and Ap1 transcription sites determines the expression of the tumor-suppressing gene *maspin.* Cell growth Diff 1997; 8(2):179-186.
12. Yamada N, Tamai Y, Miyamoto H, Nozaki M. Cloning and expression of the mouse *Pse* gene encoding a novel ets family member. Gene 2000; 241(2):267-274.
13. Zou Z, Gao C, Nagaich AK, Connell T, Saito S, Moul JW et al. p53 regulates the expression of the tumor suppressor gene *maspin.* J Biolog Chem 2000; 275(9):6051-6054.
14. de Launoit Y, Chotteau-Lelievre A, Beaudoin C, Coutte L, Netzer S, Brenner C et al. The PEA3 group of ETS-related transcription factors—Role in breast cancer metastasis. Adv Exp Med Biol 2000; 480:107-116.

15. Domann FE, Levy JP, Finch JS, Bowden GT. Constitutive Ap-1 DNA-binding and transactivating ability of malignant but not benign mouse epidermal-cells. Molec Carcinog 1994; 9(2):61-66.
16. Bowden GT, Schneider B, Domann R, Kulesz-Martin M. Oncogene activation and tumor-suppressor gene inactivation during multistage mouse skin carcinogenesis. Cancer Res 1994; 54(7):S1882-S1885.
17. Smith LM, Birrer MJ, Stampfer MR, Brown PH. Breast cancer cells have lower activating protein 1 transcription factor activity than normal mammary epithelial cells. Cancer Res 1997; 57(14):3046-3054.
18. Gardiner-Garden M, Frommer M. CpG islands in vertebrate genomes. J Mol Biol 1987; 196(2):261-282.
19. Baylin SB, Herman JG, Graff JR, Vertino PM, Issa JP. Alterations in DNA methylation: A fundamental aspect of neoplasia. Adv Cancer Res 1998; 72:141-196.
20. Clark SJ, Harrison J, Paul CL, Frommer M. High sensitivity mapping of methylated cytosines. Nucleic Acids Res 1994; 22(15):2990-2997.
21. Wade PA, Gegonne A, Jones PL, Ballestar E, Aubry F, Wolffe AP. Mi-2 complex couples DNA methylation to chromatin remodelling and histone deacetylation. Nature Genetics 1999; 23(1):62-66.
22. Pfeifer GP, Steigerwald SD, Mueller PR, Wold B, Riggs AD. Genomic sequencing and methylation analysis by ligation mediated Pcr. Science 1989; 246(4931):810-813.
23. Jiang WG, Hiscox S, Horrobin DF, Bryce RP, Mansel RE. Gamma linolenic acid regulates expression of maspin and the motility of cancer cells. Biochem Biophys Res Comm 1997; 237(3):639-644.
24. Li JJ, Colburn NH, Oberley LW, *Maspin* gene expression in tumor suppression induced by overexpressing manganese-containing superoxide dismutase cDNA in human breast cancer cells. Carcinogenesis 1998; 19(5):833-839.
25. Tsuji T, Usui S, Aida T, Tachikawa T, Hu GF, Sasaki A et al. Induction of epithelial differentiation and DNA demethylation in hamster malignant oral keratinocyte by ornithine decarboxylase antizyme. Oncogene 2001; 20(1):24-33.
26. Jones PL, Veenstra GJC, Wade PA, Vermaak D, Kass SU, Landsberger N et al. Methylated DNA and MeCP2 recruit histone deacetylase to repress transcription. Nat Gen 1998; 19(2):187-191.
27. Bird A, Taggart M, Frommer M, Miller OJ, Macleod D. A fraction of the mouse genome that is derived from islands of nonmethylated, CpG-rich DNA. Cell 1985; 40(1):91-99.
28. Cooper DN, Taggart MH, Bird AP. Unmethylated domains in vertebrate DNA. Nucleic Acids Res 1983; 11(3):647-658.
29. Spring P, Nakashima T, Frederick M, Henderson Y, Clayman G. Identification and cDNA cloning of headpin, a novel differentially expressed serpin that maps to chromosome 18q. Biochem Biophys Res Comm 1999; 264(1):299-304.
30. Ozaki K, Nagata M, Suzuki M, Fujiwara T, Miyoshi Y, Ishikawa O et al. Isolation and characterization of a novel human pancreas-specific gene, pancpin, that is down-regulated in pancreatic cancer cells. Genes Chromosomes Cancer 1998; 22(3):179-185.

Chapter 7

Maspin Suppresses Breast Cancer Cell Invasiveness by Modulating Integrin Expression and Function

Richard E.B. Seftor, Valerie A. Odero, Elisabeth A. Seftor and Mary J.C. Hendrix

Introduction

Although the novel tumor suppressor gene *maspin* (mammary serine protease inhibitor) was originally isolated from normal mammary epithelium by subtractive hybridization and differential display almost seven years ago,[1-2] it is still unclear how it functions molecularly and biologically to regulate tumor cell motility, invasion and metastasis.[1-3] The maspin protein has an M_r of 42,000 and contains sequence homology with members of the serine protease inhibitor superfamily (serpins), including plasminogen activator inhibitor-1, -2 (PAI-1 and PAI-2) and α1-antitrypsin, as well as sequence homology with noninhibitor serpins, such as ovalbumin.[3] This apparent dual nature of maspin is consistent with the observations that while recombinant maspin can act at the cell membrane to inhibit cell migration and invasion and requires an intact reactive site loop, it can also function as a substrate rather than an inhibitor for a number of different serine proteases (e.g., tissue- and urokinase-type plasminogen activators; tPA and uPA).[3-8] When acting as a serine protease inhibitor in vitro, maspin binds specifically to purified single chain tPA (sctPA) activated to cleave plasminogen to plasmin and results in biphasic effects on sctPA. This suggests a complex interaction between maspin and sctPA which could contribute to the regulation of plasminogen activation by sctPA when bound to the epithelial cell surface.[9] Recently, recombinant maspin was shown to inhibit plasminogen activation to plasmin associated with uPA activity (but not tPA activity) at the cell surface of the prostate carcinoma cell line DU-145. This correlated quantitatively with maspin's inhibition of cell motility in vitro.[10] There has also been a recent report that maspin is regulated by p53 in breast and prostate cancer cells lines.[11, 12] This suggests that maspin and p53 cooperate in the negative regulation of tumor cell invasion and metastasis. Taken together, these observations indicate that maspin may play a significant role in regulating processes that are associated with the progression and metastatic cascade of

Maspin, edited by Mary J.C. Hendrix. ©2002 Eurekah.com

certain cancers (e.g., breast and prostate cancer), and could thereby present an unique and specific target for the diagnosis and therapeutic intervention of these cancers.

Tumor Cell Invasion and Metastasis

Tumor cell metastasis is a complex process which involves tumor cells moving from their primary site of growth and proliferation to distant sites in the body. The foci formed at these sites may act to harbor tumor cells that can remain dormant for extended periods of time (as in certain uveal metastatic melanoma),[13] or function as sites of renewed tumor cell growth, proliferation and metastasis. As a result of the inherent complexity of this cascade, it has been helpful to describe and examine this process in terms of a continuum where discrete, specific steps may be identified, modeled and studied.[14,15] In order to disseminate throughout the body, tumor cells must leave the primary area of growth and get in (intravasate) and out (extravasate) of the blood and lymphatic systems, their primary routes for dispersal. During this process, tumor cells must interact with a changing environment, avoid host defenses, and recognize and lodge at sites that can facilitate and support their renewed growth and proliferation. A key component of the metastatic cascade concerns how tumor cells leave the primary tumor and invade the vasculature. In this regard, tumor cell invasion has been defined as a series of steps which describe: a) how tumor cells attach to and interact with their extracellular environment (both cell-cell and cell-extracellular matrix); b) what proteases are involved in the proteolytic digestion of the extracellular matrix and how these proteases are regulated; and c) how tumor cells move through the digested barrier.[14-17]

Integrins and Their Role in Tumor Cell Invasion

A primary focus of studies concerned with cancer progression and metastasis is the examination of how cells interact with their different, and at times rapidly changing, environment. In this regard, the family of transmembrane glycoproteins called integrins play a key role(s) in the invasive and metastatic processes. Integrins are heterodimeric glycoproteins composed of non-covalently linked α and β subunits which contain a short transmembrane segment, carboxy-terminal cytoplasmic domains of variable size, and large, extracellular ligand-binding sites composed of the N-terminal domains of the α and β subunits. In general, ligand specificity of the integrin is determined by the α subunit, and ligand-specific signals are conveyed to the cell via the cytoplasmic tail of some α subunits. In contrast, ligand-independent clustering of integrins to focal adhesion sites, where they become organized at the ends of actin filaments and associate with the proteins vinculin, talin, and α-actinin, occurs through the cytoplasmic tail of some β subunits. While just over 20 integrins have been identified so far, the theoretical number of possible integrins (including spliced variants) is greater than 100. Integrins historically fall into three groups based on similar chain structures and/or ability to recognize similar protein or adhesion motifs. These groups include the β_1-containing integrins, β_2-containing integrins and β_3- (α_v)-containing integrins. While some α subunits associate with only one of the different β subunits, other α subunits associate with more than one β subunit. The most promiscuous subunit, α_v, has been found associated with at least five different β subunits in various different cell lines (for review see Refs. 18, 19).

Classically, integrins are cell adhesion molecules involved in both cell-extracellular matrix and cell-cell adhesion interactions. However, subsequent studies have shown that integrins also play a significant role in signal transduction events, gene expression, cell proliferation, regulation of cell apoptosis (and anoikis), invasion and metastasis, embryogenesis, tumor progression, inflammation and immunity, hemostasis, angiogenesis, and in mediating the entry of certain infectious agents into cells.[18,19] While cells can vary in their ability to respond to different environmental cues as a result of the particular integrin(s) expressed on their surface, they may also respond differently to specific ligands at different times and under different conditions through activation and deactivation of their integrins. This change in activation state involves conformational changes in the integrins' extracellular domain and can result from alterations in the degree of phosphorylation of the β subunit,[20] as well as interaction between integrins and lipid-derived mediators.[21]

Another parameter contributing to the modulation and function of integrins is their association and interaction with other integrins, as well as other membrane proteins.[22-31] Cells can express a number of different integrins on their surface while at the same time experience a complex extracellular environment which contains many different integrin ligands. While each ligand has the potential to interact with a different integrin, there are individual ligands that can interact with a number of different individual integrins, and individual integrins that can interact with a number of different ligands. In order for cells to respond in a coordinated manner to this complex (and potentially changing) environment, recent studies suggest that there is a unidirectional cascade between transducer integrins, which can modulate the function of other integrins, and integrins (i.e., target integrins) which are subsequently modulated by these transducer integrins.[28] This coordination and modulation of an intgerin's function by another integrin is called integrin crosstalk,[28-31] and appears to occur by way of the integrins' β-subunit cytoplasmic tails.[30]

An important example of how membrane associated proteins other than integrins modulate integrin function has been observed using the human embryonic kidney 293 cell model where the cells were transfected with cDNA for the urokinase-type plasminogen activator receptor (uPAR).[23] uPAR is the major cellular binding site for urokinase-type plasminogen activator (uPA) which operates as a fibrin-independent, largely receptor-bound plasminogen activator,[32,33] and the stability of uPA bound to uPAR is directly coupled to the activity of uPA.[32] Plasminogen activator inhibitor-1 (PAI-1) can bind active uPA to form a complex which can bind the α_2-macroglobulin/lipoprotein-receptor related protein (LRP) and result in the clearance of uPA/PAI-1 and uPAR from the cell surface. This suggests that PAI-1 regulates both fibrinolysis and the cell-surface expression of uPAR. When the human embryonic kidney 293 cells were transfected with uPAR, the resulting glycosylphosphatidylinositol (GPI)-linked cell surface uPAR was found to interact with the active form of β_1-containing integrins and caveolin to form a stable integrin-uPAR-caveolin complex. Although this complex was found to suppress normal β_1-integrin binding to fibronectin, uPAR itself can function as an adhesion receptor for vitronectin and contains distinct sites for binding both vitronectin and urokinase. As a result, transfection of uPAR into these cells resulted in a marked increase in their ability to adhere to vitronectin, and an inhibition of their normal β_1-dependent adhesion to fibronectin.[23] This observation, however, contrasts with other reports that indicate that uPAR promotes normal functions in β_1- and β_2-containing integrins.[33] A possible explanation for this paradox may be found in recent reports that indicate that integrin clusters contain only a few uPAR molecules that are bound to the integrins in a ligand-like fashion.[23,33] The uPAR functions by enriching the clusters with signaling molecules while at the same time preventing a few uPAR-bound integrins from binding their natural ligands. This results in a sacrifice of some ligand-binding capacity for an enrichment of integrin clusters with signaling molecules.[23,33]

Classically, the ability of cells to recognize, attach and adhere to specific extracellular components is a primary function of integrins; however, they have also been shown to play a significant role in transducing signals to the cell that contribute to regulating the response and interactions of the cells to their extracellular environment. Work by Werb and colleagues[34] demonstrated that perturbation or ligation of the $\alpha_5\beta_1$ integrin (classical fibronectin receptor) with fibronectin fragments could transduce extracellular signals in rabbit synovial fibroblasts which resulted in a change in the expression and extracellular levels of collagenase and stromelysin. Furthermore, subsequent work demonstrated that ligation of the $\alpha_v\beta_3$ integrin (classical vitronectin receptor) on a moderately invasive human melanoma cell line (A375M) with vitronectin (either bound or soluble), or an activating antibody to the α_v subunit, increased the expression and extracellular levels of the matrix-metalloproteinase-2 (MMP-2) concomitant with an increase in the cells' ability to invade in vitro.[35] In contrast, however, it was found that a highly aggressive human melanoma cell line (C8161), which expressed little-to-no $\alpha_v\beta_3$ integrin on the cell surface, did not respond in the same way to either vitronectin or the activating anti-α_v antibody. These cells became more invasive in vitro when treated with either fibronectin, an anti-$\alpha_5\beta_1$ perturbing antibody, or an anti-α_5 integrin subunit antibody. These changes were accompanied by increases in the expression and extracellular levels of MMP-2.[36] Together, these results suggested that integrin interactions could result in the generation of signals which contribute to changes in cellular behavior and pathological phenotype.

rMaspin Suppresses Breast Cancer Cell Invasiveness In Vitro[22]

Although the role of maspin as an inhibitory or non-inhibitory serpin remains unclear, we examined how maspin might function in modulating tumor cell aggressiveness by treating the human breast cancer cell line MDA-MB-435 with recombinant maspin (rMaspin), then measuring potential changes in their invasive ability in vitro using the Membrane Invasion Culture System (MICS) model.[35-37] Treatment of these cells with rMaspin (20 µg/ml for 24 hours) resulted in a 43% decrease in their ability to invade through a fibronectin/gelatin defined matrix barrier in the MICS assay. Based on previous observations that maspin acts at the cell surface, we evaluated the integrin profiles of MDA-MB-435 cells before and after treatment with rMaspin. Using fluorescence activated cell (FAC) analyses, we determined the percent mean fluorescence of 5,000 cells labeled with the appropriate primary anti-integrin subunit antibody and FITC-conjugated secondary antibody, and corrected for autofluorescence and nonspecific binding by the secondary antibody. We compared the resulting percent mean fluorescence of the treated and untreated cells and found that there was a significant increase in expression of the α_3- and α_5-containing integrins (21% and 49%, respectively), a modest increase in the α_4-containing integrin (7%), and decreases in the α_2-, α_6-, α_v- and β_1-containing integrins (11%, 15%, 23% and

21%, respectively) in response to treatment with rMaspin. Northern blot analysis corroborated the FAC analysis for the α_5-integrin subunit and demonstrated a 30% increase in the α_5-integrin subunit mRNA in the rMaspin-treated cells, although no change was seen in the mRNA for the α_3 subunit. Furthermore, immunofluorescence microscopy demonstrated an increase in distribution of the $\alpha_5\beta_1$ integrin (classical fibronectin receptor) on the rMaspin treated cells, and a change in cell morphology to a more epithelial-like phenotype compared to the more fibroblastic phenotype displayed by the control cells.[22]

To determine whether the rMaspin-induced change in integrin profile was biologically significant, cellular adhesion and in vitro invasion assays were performed on the control compared to the rMaspin treated cells. Cells treated with rMaspin were 27% more adhesive to fibronectin than the untreated control cells, but showed little-to-no change in adhesion to laminin, vitronectin, collagen IV or collagen I. Furthermore, pretreating the rMaspin-treated cells with a function blocking antibody to the $\alpha_5\beta_1$ integrin prior to plating in the adhesion assay prevented the induced increase in adhesion. Cells treated with rMaspin were also 43% less invasive (in vitro) through a fibronectin/gelatin matrix compared to the untreated control cells. Cells pretreated with a function blocking antibody to the $\alpha_5\beta_1$ integrin for 15 minutes before the assay resulted in a recovery of the cells' invasive potential to within 9% of the control cells' invasiveness. As a control for the specificity of the $\alpha_5\beta_1$/rMaspin interaction, the invasion assay was repeated using a laminin/gelatin matrix barrier with and without a blocking antibody to the α_6-integrin subunit. Cells treated with rMaspin were 27% less invasive through the laminin/gelatin matrix than the untreated control cells, and addition of a blocking anti-α_6 antibody did not change this result.

While maspin occurs in normal breast tissue and normal breast cells in culture, it is not present in breast cancer tissue and breast cancer cells in culture. The results presented here suggest a biologically relevant explanation concerning the mechanism(s) by which addition of rMaspin inhibits the maspin-deficient human breast cancer cell line MDA-MB-435 from invading in vitro. Addition of rMaspin to these cells resulted in a decrease in their in vitro invasiveness coincident with an increase in their cell surface expression of the α_5-containing integrin, and an increase in their adherence to fibronectin. These results were corroborated by Northern blot analysis, which showed an increase in mRNA for the α_5-integrin subunit in rMaspin treated cells, and by the observations that addition of a blocking antibody to the $\alpha_5\beta_1$ integrin 1) inhibited their increased adhesion to fibronectin, and 2) facilitated their ability to invade through the fibronectin/gelatin matrix in vitro at a rate equivalent to the untreated control cells. These observations also support previous observations that there is a competitive reversal of rMaspin action in breast cancer cells treated with an RGD peptide known to block integrin function.[37] Together, these results indicate a functional change in the response of MDA-MB-435 cells to their environment which can be induced by rMaspin and involves both transcriptional and translational processes in the cells. A significant observation from these studies is that the fibroblastic-like, invasive and metastatic cells appear to assume a more benign, epithelial-like morphology in response to rMaspin.[22] While E-cadherin expression did not appear to increase in response to rMaspin, this could indicate that the actual junctional adhesion complexes had not formed during the 24 hour period of these observations. Although rMaspin treatment also caused a decrease in the cell surface expression of the α_2-, α_6-, α_v- and β_1-containing integrins, a change was not seen in the cells' ability to adhere to any of the

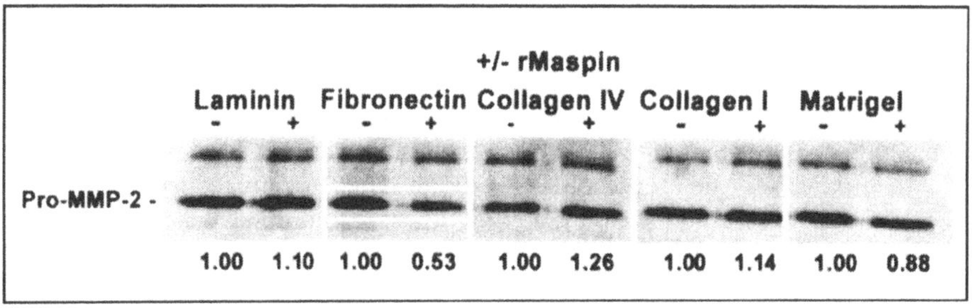

Fig. 1. rMaspin reduces the extracellular levels of pro-MMP-2 in MDA-MB-435 cells plated on fibronectin. MDA-MB-435 cells were plated in serum-free medium onto laminin, fibronectin, collagen IV, collagen I or Matrigel matrices plus or minus 20 μg/ml of rMaspin. After 24 hours, the media were collected, centrifuged to remove cells and debris, and the supernatants analyzed by substrate incorporated sodium dodecylsulfate polyacrylamide gel electrophoresis (i.e., SDS-PAGE zymography). After electrophoresis, the gels were stained, destained, and images of the gels captured digitally. These images were then analyzed for the relative pixel densities of the bands in the rMaspin treated lanes compared to the untreated control lanes normalized to 1.00. A white box highlights the results for the fibronectin samples which demonstrate the greatest difference in extracellular levels of MMP-2 in response to treatment with rMaspin.

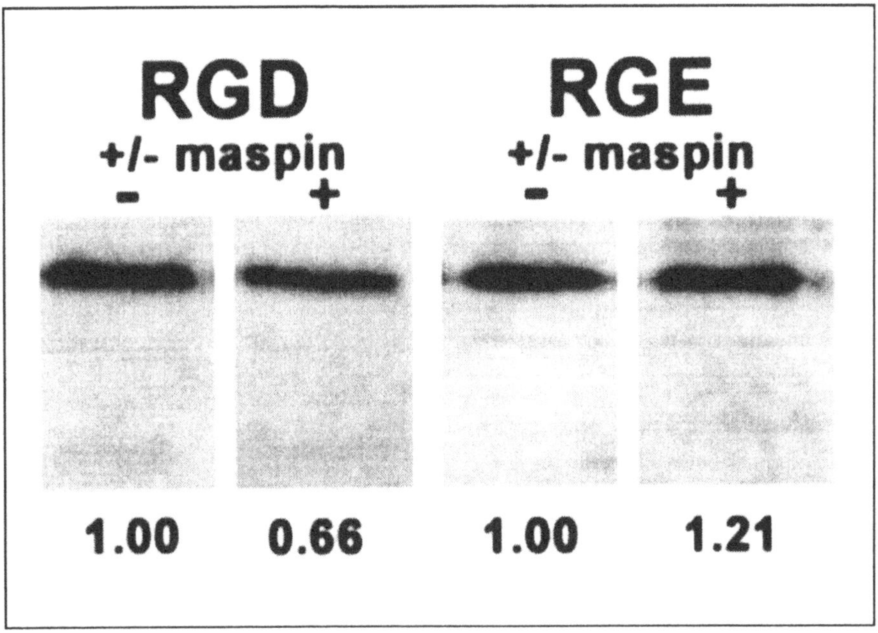

Fig. 2. An RGD peptide abrogates the decrease in MMP-2 activity induced by rMaspin in MDA-MB-435 cells cultured on fibronectin. MDA-MB-435 cells were plated in serum-free medium on fibronectin, plus or minus rMaspin, plus or minus RGD or RGE peptides. After 24 hours, the media were collected, centrifuged to remove cells and debris, and the supernatants run on 10% SDS-PAGE zymograms. Images were obtained and analyzed as described in Figure 1.

ligands normally associated with these integrins. Furthermore, the inability of a blocking antibody to the α_6-subunit to restore the invasiveness of the rMaspin treated cells through the laminin/gelatin matrix barrier in the in vitro invasion assay suggests that there is a specific relationship between the α_5-containing integrin and rMaspin which apparently does not exist with the other integrins.

In summation, these observations indicate that rMaspin reduces the invasive potential of MDA-MB-435 cells by altering their integrin profile which changes how they perceive and interact with their extracellular environment. rMaspin treated cells become more adherent to fibronectin-containing biological substrates and are subject to phenotypic changes that result in the conversion from an invasive, fibroblastic phenotype to an epithelial-like, less invasive phenotype.

Role of MMP-2 in the Suppression of Breast Cancer Cell Invasiveness In Vitro by Maspin

As described above, MDA-MB-435 cells treated with rMaspin were less invasive in vitro, more adherent to fibronectin, and expressed a significant increase in the a_5-containing integrin on their cell surface. Given our previous work which demonstrated a change in the expression and extracellular levels of MMP-2 in response to perturbation and ligation of different integrins on human melanoma cells coincident with the cells' ability to invade in vitro,[35,36] we examined whether treatment with rMaspin altered the expression and extracellular levels of MMP-2 in MDA-MB-435 cells. Cells plus or minus 20 μg/ml rMaspin were plated in serum-free medium on laminin, fibronectin, collagen IV, collagen I or Matrigel. After 24 hours, the media was removed from the cells, centrifuged to remove cells and cellular debris, and the supernatants analyzed by substrate incorporated sodium dodecylsulfate polyacrylamide gel electrophoresis (i.e., SDS-PAGE zymography). The resulting zymograms were digitized, and the images analyzed to determine pixel densities of the bands relative to the control samples normalized to a value of 1.00. As shown in Figure 1, there was a 47% decrease in MMP-2 (pro-enzyme) activity in the sample plated on fibronectin and treated with rMaspin, and relatively little change in any of the other samples. Northern blot analysis corroborated this result and revealed a 35% decrease in mRNA for MMP-2 in the cells treated with rMaspin and plated on fibronectin. In order to determine if the $\alpha_5\beta_1$ integrin is involved in modulating the expression of MMP-2, this experiment was repeated on fibronectin in the presence of either an RGD peptide known to block the function of $\alpha_5\beta_1$ and other RGD-binding integrins, or an RGE peptide (non-blocking) control peptide. As shown in Figure 2, treatment with the RGD peptide abrogated the decrease in MMP-2 activity induced by rMaspin while the RGE peptide did not. This result suggests that rMaspin is modulating the expression of MMP-2 in MDA-MB-435 cells through an integrin signaling pathway, and in light of our previous observations, most probably via the $\alpha_5\beta_1$ integrin.

Maspin Re-expression Alters Cell Morphology, uPAR and $\alpha_5\beta_1$ Integrin Distribution on Human Breast Cancer Cells

As an extension of this work, MDA-MB-435 cells were transfected with *maspin* cDNA and their morphology, in vitro invasiveness and distribution of the $\alpha_5\beta_1$ integrin and uPAR examined. Re-expression of *maspin* caused these cells to appear less fibro-

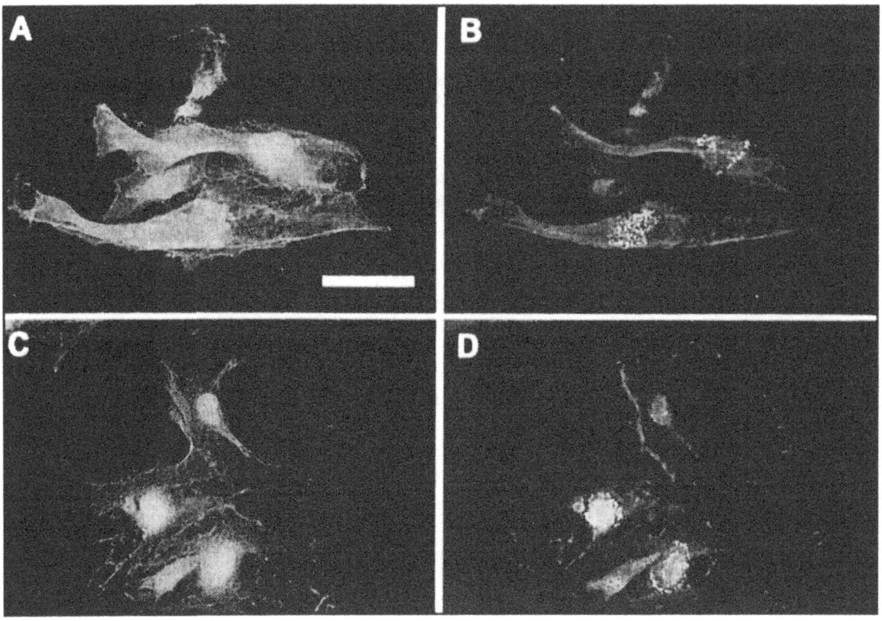

Fig. 3. Immunocytochemical analysis of Nn-3 (*neo*/sham) and T6 (*maspin*) transfected MDA-MB-435 cells using anti-$\alpha_5\beta_1$ and anti-urokinase plasminogen activator antibodies. Nn-3 (A, B) and T6 (C, D) transfected cells were cultured on glass coverslips for 24 hours, fixed in 3.7% formaldehyde in phosphate buffered saline (PBS) and probed with a polyclonal anti-$\alpha_5\beta_1$ integrin antibody and a monoclonal anti-uPAR antibody. The cells were then treated with fluorescein-conjugated (green) anti-rabbit antibody to detect the anti-$\alpha_5\beta_1$ antibody/integrin complex, and rhodamine-conjugated (red) anti-mouse antibodies to detect the anti-uPAR antibody/uPAR complex. Images were captured digitally into either the green or red channel of the digital camera from the same field of cells for each image. (Bar = 50 μm.)

blastic and assume a more epithelial-like phenotype (Fig. 3A and B sham transfected controls compared to C and D *maspin* transfectants). In addition, the $\alpha_5\beta_1$ integrin appeared widely distributed throughout the more extensive filopodial and lamillapodial network projections observed in the T6 transfectant (Fig. 3C) compared to the Nn-3 control *neo*/sham transfectants (Fig. 3A). While uPAR appeared clustered polarly in the perinuclear region of the Nn-3 cells (Fig. 3B), it displayed a uniform perinuclear distribution pattern in the T6 cells (Fig. 3D). Western blot analysis of whole cell lysates from *maspin* transfected T6 and Tn-15 cells cultured on either plastic or fibronectin confirmed that these cells expressed the maspin protein compared to the control Nn-12 sham transfected cells which did not express maspin when cultured on either surface (Fig. 4). As shown in Figure 5, T6 and Tn-15 cells were 46% and 41% less invasive in vitro (respectively) than the control *neo*/sham transfectants Nn-3 and Nn-12.

Conclusion

It is clear from the information presented in this review that maspin, whether added exogenously as a recombinant molecule or transfected into a human breast cancer cell line, can regulate a cell's invasive and morphological properties, as well as

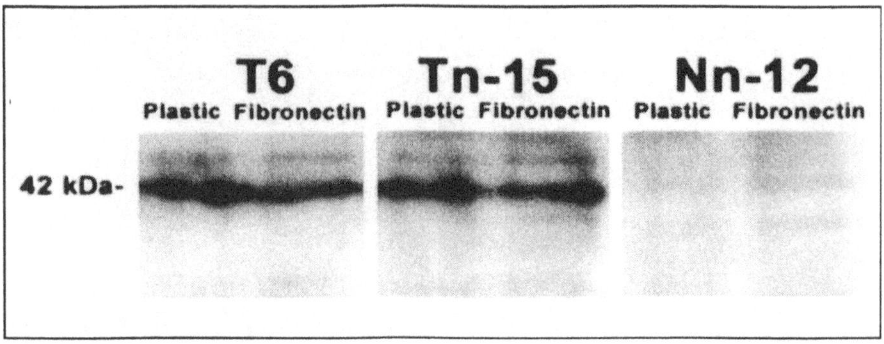

Fig. 4. MDA-MB-435 cells transfected with *maspin* express the M_r 42,000 maspin protein. The T6 and Tn-15 *maspin* transfectants and Nn-12 *neo*/sham transfected cells were plated on plastic or fibronectin for 1 hour. The cells were then scraped into RIPA buffer and equal amounts of protein loaded on 10% SDS-PAGE gels. After electrophoresis, the proteins were electrophoretically transferred to a polyvinylidene fluouride membrane and the resulting blot probed with a polyclonal anti-maspin antibody. After treating the blot with a peroxidase-conjugated secondary antibody, the maspin protein was located and identified using enhanced chemoluminescence (ECL).

the expression and extracellular levels of the invasion/metastasis associated matrix metalloproteinase MMP-2. Furthermore, this regulation appears to occur through a signaling pathway associated with the $\alpha_5\beta_1$ integrin. In light of the recent report that recombinant maspin inhibited plasminogen activation to plasmin associated with uPA (but not tPA) activity at the cell surface of the prostate carcinoma cell line DU-145,[10] and our observation that both uPAR and the $\alpha_5\beta_1$ integrin displayed an altered distribution pattern on T6 *maspin* transfected cells, these data strongly suggest that maspin's regulation of the pathogenic phenotype extends past the control of just MMP-2 expression and extracellular levels, and involves the regulation of the plasminogen activator/plasmin system as well. This is highly significant since the plasminogen activator/plasmin system is not only important for intravascular thrombolysis, but has also been implicated in the processes of angiogenesis, tumor progression and inflammatory reactions important to host defense and wound repair.[24] Collectively, these results identify unique and novel interactions that occur between maspin and breast cancer cells, and begin to reveal how these interactions function to regulate tumor cells aggressiveness. The next phase of this research will begin an examination of specific integrin and uPAR/PAR signaling pathways that might be involved in and/or facilitate the effects that maspin has on cells. As a result of this work, it is hoped that a better understanding of how maspin functions will lead to the identity of specific targets that will prove useful for the diagnosis and therapeutic intervention of breast and prostate (and possibly other) cancers.

Acknowledgements

This work was supported by the NIH/NCI grant CA75681 to MJCH.

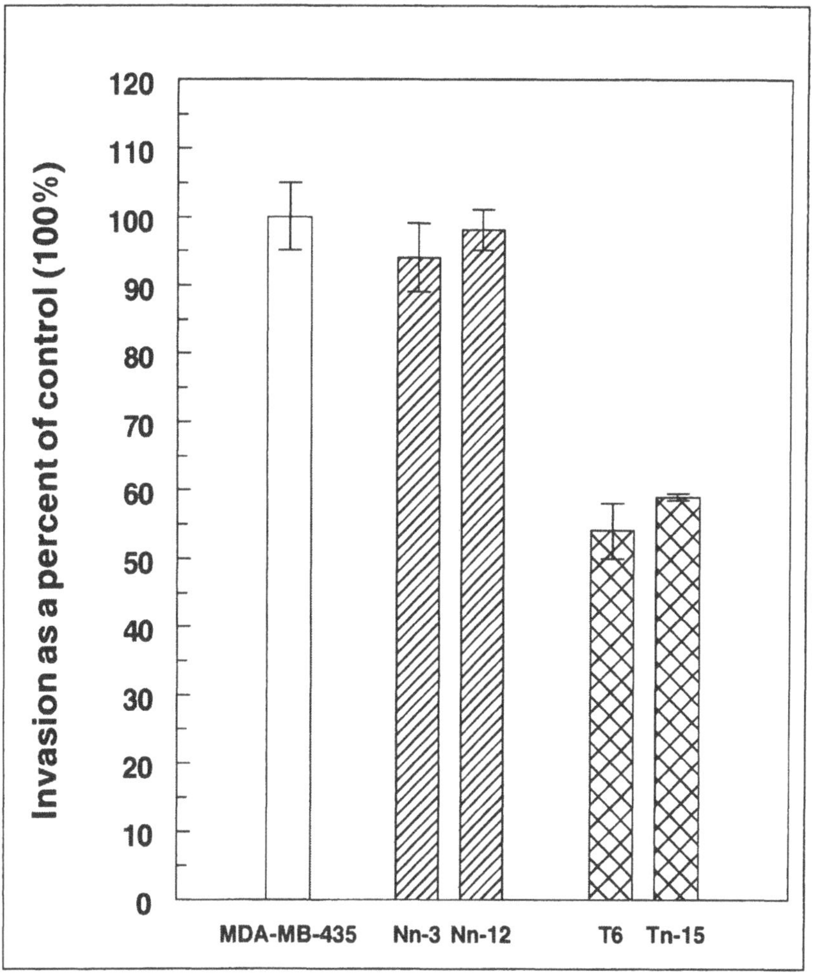

Fig. 5. MDA-MB-435 cells transfected with *maspin* are less invasive in vitro through a fibronectin/gelatin matrix than *neo*/sham transfected cells or untreated control cells after 24 hours. MDA-MB-435 untreated control cells and Nn-3 and Nn-12 *neo*/sham transfected cells demonstrate similar invasive potential in the MICS invasion assay. T6 and Tn-15 cells transfected with *maspin* are 46% and 41% (respectively) less invasive in vitro then the control (MDA-MB-435) or sham transfected (Nn-3 and Nn-12) cells.

References

1. Zhou Z, Anisowicz A, Hendrix MJC et al. Maspin, a serpin with tumor-suppressing activity in human mammary epithelial cells. Science 1994; 263:526-529.
2. Sager R, Sheng S, Pemberton P, Hendrix MJC. Maspin. A tumor suppressing serpin. In: Bunthert U, Birchmeier W, eds. Current Topics in Microbiology and Immunology 213. Berlin: Springer-Verlag, 1996:51-64.
3. Sheng S, Pemberton P, Sager R. Production, purification and characterization of recombinant maspin proteins. J Biol Chem 1994; 269:30988-30993.

4. Sager R. Function of maspin. Science 1994; 265:1893.
5. Pemberton PA, Wong DT, Gibson HL et al. The tumor suppressor maspin does not undergo the stressed to relaxed transition or inhibit trypsin-like serine proteases. J Biol Chem 1995; 270:15832-15837.
6. Fitzpatrick PA, Wong DT, Barr PJ et al. Functional implications of the modeled structure of maspin. Protein Eng 1996; 9:585-589.
7. Sheng S, Carey J, Seftor EA et al. Maspin acts at the cell membrane to inhibit invasion and motility of mammary and prostatic cancer cells. Proc Natl Acad Sci 1996; 93:11669-11674.
8. Pemberton PA, Tipton AR, Pavloff N et al. Maspin is an intracellular serpin that partitions into secretory vesicles and is present at the cell surface. J Histochem Cytochem 1997; 45:1697-1706.
9. Sheng S, Truong G, Fredrickson D et al. Tissue-type plasminogen activator is a target of the tumor suppressor gene maspin. Proc Natl Acad Sci 1998; 95:499-504.
10. McGowen R, Biliran H Jr, Sager R et al. The surface of prostate carcinoma DU145 cells mediates the inhibition of urokinase-type plasminogen activator by maspin. Cancer Res 2000; 60:4771-4778.
11. Hendrix MJC. De-mystifying the mechanism(s) of maspin. Nature Med 2000; 6:374-376.
12. Zhou Z, Gao C, Nagaich AK et al. P53 regulates the expression of the tumor suppressor gene. J Biol Chem 2000; 275:6051-6054.
13. Maniotis AJ, Folberg R, Hess A et al. Vascular channel formation by human melanoma cells in vivo and in vitro: Vasculogenic mimicry. Am J Pathol 1999; 155(3):739-752.
14. Liotta LA, Rao CN. Tumor invasion and metastasis. Monogr Patho 1986; 27:183-192.
15. Liotta LA, Rao CN, Wewer UM. Biochemical interactions of tumor cells with the basement membrane. Annu Rev Biochem 1986; 55:1037-1057.
16. Liotta LA. Tumor invasion and metastases—Role of the extracellular matrix: Rhoads Memorial Award lecture. Cancer Res 1986; 46:1-7.
17. Goldfarb Rh, Liotta LA. Proteolytic enzymes in cancer invasion and metastasis. Semin Thromb Hemost 1986; 12:294-307.
18. Seftor REB. Role of the β_3 integrin subunit in human primary melanoma progression. Multifunctional activities associated with $\alpha_v\beta_3$ expression. Commentary. Am J Pathol 1998; 153(5):1347-1351.
19. Seftor REB, Seftor EA, Hendrix MJC. Molecular role(s) for integrins in human melanoma invasion. Cancer Metastas Rev 1999; 18:359-375.
20. Blystone SD, Lindberg FP, Williams MP et al. Inducible tyrosine phosphorylation of the β_3 integrin requires the α_v integrin cytoplasmic tail. J Biol Chem 1996; 271:31458-31462.
21. Hynes RO. Integrins: Versatility, modulation, and signaling in cell adhesion. Cell 1992; 69:11-25.
22. Seftor REB, Seftor EA, Sheng S et al. *maspin* suppresses the invasive phenotype of human breast carcinoma. Cancer Res 1998; 58:5681-5685.
23. Wei Y, Lukashev M, Simon DI et al. Regulation of integrin function by the urokinase receptor. Science 1996; 273:1551-1555.
24. Chapman HA. Plasminogen activators, integrins, and coordinated regulation of cell adhesion and migration. Curr Opin Cell Biol 1997; 9:714-724.
25. Schwartz MA. Integrins, oncogenes, and anchorage independence. J Cell Biol 1997; 139:575-578.
26. Bauer JS, Schreiner CL, Giancotti FG et al. Motility of fibronectin receptor-deficient cells on fibronectin and vitronectin: collaborative interactions among integrins. J Cell Biol 1992; 116:477-487.
27. Simon KO, Nutt EM, Abraham DG et al. the $\alpha_v\beta_3$ integrin regulates $\alpha_5\beta_1$-mediated cell migration toward fibronectin. J Biol Chem 1997; 272:29380-29389.
28. Blystone SD, Slater SE, Williams MP et al. Molecular mechanism of integrin crosstalk: $\alpha_v\beta_3$ suppression of calcium/calmodulin-dependent protein kinase II regulates $\alpha_5\beta_1$ function. J Cell Biol 1999, 145:889-897.
29. Blystone SD, Graham FP, Lindberg FP et al. Integrin $\alpha_v\beta_3$ differentially regulates adhesive and phagocytic functions of the fibronectin receptor $\alpha_v\beta_1$. J Cell Biol 1994; 127:1129-1137.

30. Blystone SD, Lindberg, FP, LaFlamme SE et al. Integrin β_3 cytoplasmic tail is necessary and sufficient for regulation of $\alpha_5\beta_1$ phagocytosis by $\alpha_v\beta_3$ and integrin associated protein. J Cell Biol 1995; 130:745-754.
31. Porter JC, Hogg N. Integrin cross talk: Activation of lymphocyte function-associated antigen-1 on human T cells alters $\alpha_4\beta_1$. J Cell Biol 1997; 138:1437-1447.
32. Chapman HA. Plasminogen activators, integrins, and the coordinated regulation of cell adhesion and migration. Curr Opin Cell Bio 1997; 9:714-724.
33. Ossowski L, Aguirre-Ghiso JA. Urokinase receptor and integrin partnership: coordination of signaling for cell adhesion, migration and growth. Curr Opin Cell Biol 2000; 12:613-620.
34. Werb, Z, Tremble PM, Behrendtsen O et al. Signal transduction through the fibronectin receptor induces collagenase and stromelysin gene expression. J Cell Biol 1989; 109:877-889.
35. Seftor REB, Seftor EA, Gehlsen KR et al. Role of the $\alpha_v\beta_3$ integrin in human melanoma cell invasion. Proc Natl Acad Sci 1992; 89:1557-1561.
36. Seftor REB, Seftor EA, Stetler-Stevenson WG et al. The 72 kDa type IV collagenase is modulated via differential expression of $\alpha_v\beta_3$ and $\alpha_5\beta_1$ integrins during human melanoma cell invasion. Cancer Res 1993; 53:3411-3415.
37. Hendrix MJC, Seftor EA, Thomas PA et al. Biological function(s) of maspin. Proc Am Assoc Cancer Res 1997; 38:436A.

Chapter 8

The Role of Maspin in Tumor Progression and Normal Development

Ming Zhang

Introduction

Serine protease inhibitors (serpins) are comprised of a large family of molecules that play a variety of physiological roles in vivo.[1-3] Not all molecules that inhibit serine proteases are termed serpins. But all serpins have a very special protein structure and molecular weight about 400 amino acids.[2] Serpins exist in almost every organism, from virus to mammals.[3-5] They can be divided into two categories: inhibitory and non-inhibitory serpins. Non-inhibitory serpins, typified by ovalbumin and PEDF, do not exhibit protease inhibitor activity, but rather function as a storage protein and neural differentiation factor, respectively.[6,7] Inhibitory serpins ablate serine proteases through their functional domain-reactive site loop (RSL).[2] Interestingly, some inhibitory serpins have evolved other regulatory functions. For example, plasminogen activator inhibitor 1 (PAI-1) not only specifically inhibits tPA and uPA,[8,9] but also regulates cell adhesion, which is independent of its protease inhibitor function, by blocking integrin $\alpha_v\beta_3$ binding to vitronectin.[10,11] This implies that serpins not only play diverse roles as a class, but also a single serpin molecule may possess multiple functions.[12]

Maspin is a unique member of the serpin family that shares extensive homology with monocyte-neutrophil elastase inhibitor,[3] PAI-2 and other serpins.[13,14] Initially identified as a class II tumor suppressor gene, maspin has been shown to inhibit invasion and motility of mammary tumors.[13,15,16] Tumor transfectants expressing maspin exhibit decreased growth and metastasis in nude mice.[13] Maspin gene expression is not detected in most breast tumors and loss of its expression is correlated with tumor invasiveness.[17] Maspin is also found to be a potent angiogenesis inhibitor.[18] In human breast tissue, maspin seems to be present more in luminal than myoepithelial cells, and it has been suggested that those maspin-expressing myoepithelial cells form a defensive barrier for the progression from ductal carcinoma in situ to more invasive carcinomas.[19] A dispute exists regarding whether maspin acts as a protease inhibitor among biochemists.[20-22] Evidence from my laboratory indicates that maspin, regardless of whether it inhibits protease or not, possesses other functions independent of anti-protease action(Zhang et al, unpublished data).[16]

In this Chapter I intend to summarize the studies on maspin done by my group in the last few years and some works that are done by others but are closely related to ours. They will be divided into three sections: 1) maspin gene expression; 2) role of maspin in normal mouse development and 3) role of maspin in tumor progression and angiogenesis.

Maspin Gene Expression

Maspin Gene and Chromosome Localization

Human maspin encodes a 375 amino acid protein with a molecular weight of 42 k dalton. By FISH in situ analysis maspin is located at chromosome 18q21.3.[13] To determine maspin gene structure, we have cloned maspin genomic DNA and sequenced all exon/intron borders as well as partial introns. Our data show that maspin gene contains 7 exons and 6 introns with a total length of 28 kb (Zhang et al, unpublished). It contains a non-coding exon 1 and an unusually long intron 1 (7.4 kb). Further genetic study and computer search of recently published human genome sequence shows that this region has a cluster of serpins belonging to the ovualbumin serpin family (Clan B). These include from the direction of telomere to centromere: PI8, PI10, PAI-2, megsin, SCCA1, SCCA2, headpin, and maspin.[23,24] The gene structure of maspin is different from that of PAI2, SCCA1 and 2 in the same cluster since they all contain 8 exons and 7 introns. In addition, another cluster of serpins within the Clan B family exists in chromosome 6p25. Evidence from genetic study shows that these two clusters may have evolved from both intra- and inter-chromosome duplication.[25]

The phylogenetic tree shows that maspin is conserved with the ovalbumin family of serpins. The degree of homology at the amino acid level is : 40%, human monocyte-neutrophil elastase inhibitor (ELANH2), 39%, human PI10, 39%, Human PI8, 38%, human PI6, and 35% for human PAI-2.[23,26-28] ELANH2 and PI6 are located at chromosome 6p25.[29] Other serpins at 18q21.3 such as PI8 are also more homologous to PI9 at 6p25, supporting the hypothesis of interchromosome duplication.

Differential Expression of Maspin in Breast Normal and Tumor Cells

The maspin gene was originally isolated from normal mammary epithelial cells. Northern blot analysis had been used to evaluate the expression pattern of maspin in human cell lines. Maspin was found to be highly expressed in 70N and 76N normal mammary epithelial cells, down-regulated in 21NT and 21PT primary breast tumors, and silent in a series of metastatic tumor cells. The gene was not expressed in cells of non-epithelial origin, such as mammary fibroblast cells, foreskin fibroblast cells, and U937 human monocytic cells. Mouse homolog of maspin was also expressed highly in mouse normal mammary epithelial cells such as NMuMG, HC11. The level of expression was down-regulated in low invasive mouse mammary tumor cells and lost in a highly invasive tumor cell line.

Transcriptional Regulation of Maspin in Breast Cells

To study maspin gene regulation in breast normal and cancer cells, we cloned maspin gene promoter. The cloning of maspin promoter has been quite a challenge because of the presence of a long intron 1. Most lambda genomic clones contain about a 20 kb genomic insert. Thus, using the full length of maspin cDNA as a probe, we were unable to recover any clones containing maspin promoter since the region from exon 2 to the last exon spanned over 20 kb. Finally, a YAC clone containing a cluster of serpins including maspin was used to screen maspin genomic DNA using a 50 mer oligomer as a probe. A positive clone was identified containing a 1.2 kb 5'-flanking region, a 7.4 kb intron 1 and partial exon 2 sequence. Sequencing and primer extension analysis showed that maspin did not have a typical TATA box in its promoter, a common feature for housekeeping or ubiquitously expressed genes. The major transcription start site was located 207 bp from first ATG site and the ATG was located in the second exon.[17]

Several important transcription factor binding sites, Ets, Ap1, HRE and p53, are located within the 1 kb promoter region. To identify the responsive elements, promoter deletion and CAT assay were used to identify the activity of maspin in normal mammary epithelial cells (70N), primary tumor cells (21NT), and metastatic tumor cells (MDA-MB231) (Fig.1). As shown in Fig.1, a 1 kb upstream region was sufficient for activating transcription of maspin in normal breast cells. The activity was decreased in 21NT cells, and no detectable CAT activity was found in MDA-MB231 cells. Deletion analysis showed that deletion of an ets site located at the region from-112 bp to -90 bp, abolished completely maspin activity, suggesting that this ets was the major positive *cis* element within 1 kb responsible for up-regulation of maspin in normal mammary epithelial cells. Further analysis confirmed that this ets site cooperated with a downstream Ap1 site for synergistic activation in normal breast cells; however this cooperative transactivation between ets and Ap1 was lost in primary breast tumor 21NT cells.

Maspin Expression in Prostate Cells

Prostate cancer is the most common cancer in men. The prostate gland depends on androgenic hormones for its growth and development, analogous to the role of mammary hormones in development and morphology changes in the mammary gland.[30] The molecular events leading to the development of prostate cancer may be similar to those in breast cancer. Because maspin functions as a tumor suppressor in the mammary gland, we have examined whether maspin could play a similar tumor-suppressing role in the prostate, and more importantly, what is the mechanism underlying gene regulation of maspin in prostate cells.

Northern blot analysis was carried out with RNAs from several human normal prostate and tumor cell lines. Maspin is highly expressed in CF3, CF91, and MLC normal prostate epithelial cells, and down-regulated in LNCaP, PC3, and DU145 prostate tumors. This expression pattern is similar to the findings in the normal mammary epithelial cells and carcinomas, indicating that the down-regulation of maspin expression is a common phenotype of both breast and prostate tumors.[13,17] A 1 kb maspin promoter is active in CF3 normal prostate cells, but little activity is present in LNCaP prostate cancers. As in mammary epithelial cells, the same ets site is primarily responsible for transcriptional action of maspin in CF3 cells. In addition, we have identified an HRE site located between -297 bp and -265 bp that represses

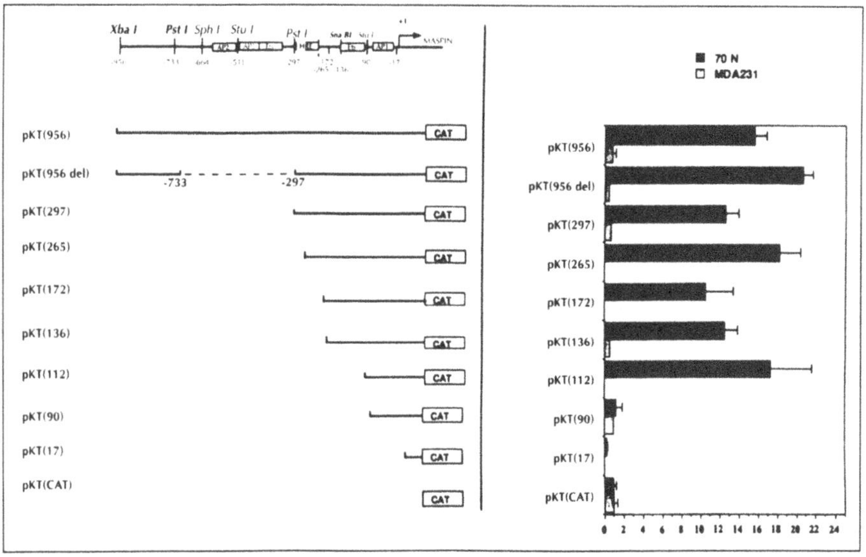

Fig. 1. Maspin CAT constructs and CAT assays. On the left: CAT constructs and a schematic representation of maspin promoter with putative transcription factor binding sites. On the right: CAT constructs were transfected into 70N and MDA-MB231 cells. Relative activity was represented by normalizing to pKTCAT. Error bars are from at least four eperiments.

transcriptional activity of maspin. Point mutations at this HRE site abolished the repression completely (see Fig.2.). The consensus HRE element has a sequence 5'-GGTACANNNTGT(T/C)CT-3'.[31] This sequence can be recognized by multiple steroid receptors, such as glucocorticoid receptor, androgen receptor, and progesterone receptor.[32] Previous studies have demonstrated that steroid receptors binding to HRE could mediate both transcription activation and repression.[33,34] The HRE site (5'-GTACTCTGATCTCC-3') in the maspin promoter is unique in that its sequence does not share very good homology with the consensus sequence. This maspin HRE sequence acted as a general transcription repressor when it was placed in front of other heterologous promoters. Furthermore, we demonstrated that repression was mediated through AR using electrophoresis mobility shift assay. Binding of AR to maspin HRE is specific and can be blocked by AR antibody.[35] These data may explain castration-mediated suppression of prostate cancer progression since ablation of androgen by castration may up-regulate maspin in prostate. From the therapeutic point of view, re-expression of maspin in the prostate tumors offers great hope for reversing the tumor phenotypes. Re-expression may be achieved by targeting both activation and repression modes. While it will likely be difficult to restore transcriptional activation of maspin through the ets site, it may be more feasible to block the repression mediated by the AR binding HRE element. Treating tumors with ligands that block the binding of AR to the HRE or with reagents that compete strongly for binding to AR are possible methods of blocking HRE mediated repression. Our discovery of HRE-mediated repression offers another opportunity to increase the expression of maspin in prostate tumors, which may in turn reduce the progressiveness of prostate cancer.

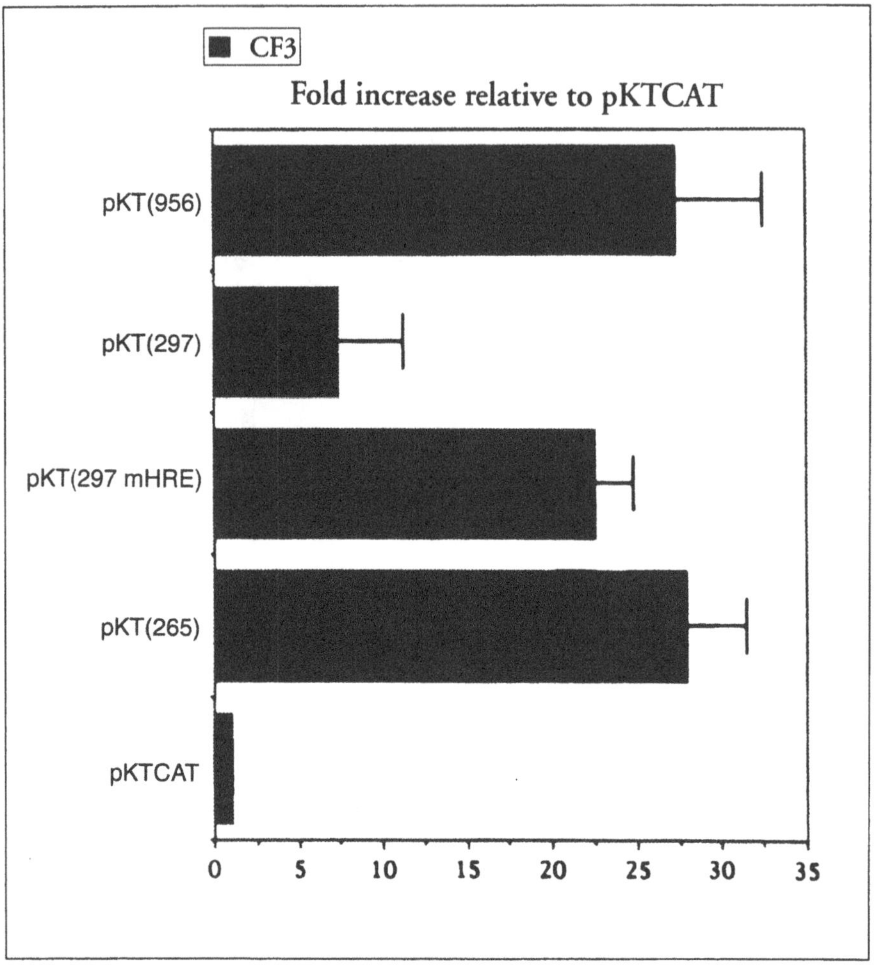

Fig. 2a. HRE element negatively regulates maspin promoter activity. Effect of mutation at HRE site on the promoter activity. CAT constructs were transfected to CF3 cells and activity normalized to pKTCAT control. Values are btained from at least three repeated experiments. Error bars are standard errors.

Regulation of Maspin by Other Transcription Factors and Reagents

A number of groups have contributed to the discovery of new transcription factors and reagents that regulate maspin expression. Notably, a recent report has demonstrated that p53 activates maspin in prostate and breast cancer cells.[36] Infection of these cancer cells through adenovirus p53 dramatically induced maspin expression with a rapid kinetics, similar to that of another well-known p53 regulated gene, p21. Inducers of p53 such as DNA-damaging reagents and drugs also induced maspin expression. They further showed that a consensus sequence for p53 binding is located in the maspin promoter and p53 binds to this site in gel shifting assay. A promoter luciferase reporter containing the p53 site was active when cotransfected with wildtype

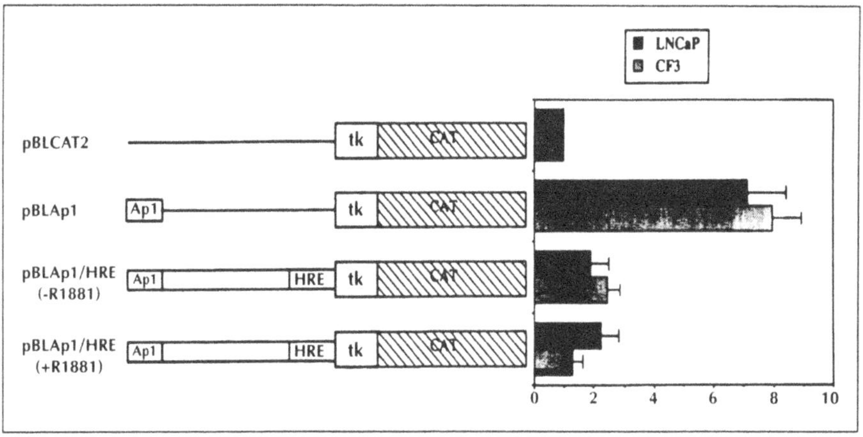

Fig. 2b. HRE element negatively regulates maspin promoter activity. Effect of HRE on the promoter activity of pBLAP1. CAT constructs were transfected to CF3 cells and activity normalized to pBLCAT2 control. Transfected cells were treated with R1881 (50nM) or vehicle for 48 hrs. Values are obtained from at least three repeated eperiments. Error bars are standard errors.

p53 into the prostate cancer cells. This finding is significant since it marks maspin as the first p53 regulated gene involved in cancer invasion and metastasis.

PPAR-γ plays an important role in adipocyte differentiation. A recent study also demonstrated that activation of PPAR-γ by its ligands caused dramatic morphological and nuclear changes that were characteristic of a more differentiated, less malignant state in breast cancer cells. Associated with these changes were the increase of maspin expression and arrest of cell cycle.[37] Thus, PPAR-γ mediated terminal differentiation of cancer cells might result from the increased maspin expression by a mechanism of transcriptional up-regulation. Gamma linolenic acid, an essential fatty acid, has anti-cancer properties. It was reported that gamma linoneic acid could induce maspin expression very rapidly.[38] Cancer cells treated with GLA were shown to have reduced spreading and migration as monitored by video-microscope, and this reduction was reversed if the cells were treated with anti-maspin antibody. Thus, the anticancer effect of GLA was thought to result from the activation of maspin expression. Similarly, maspin was found to be induced by superoxide dismutase (MnSOD) and this induction might explain the tumor suppressive effect of MnSOD in breast cancer cells.[39]

While it is relatively easy to identify the transcription factors and their regulatory sites responsible for maspin expression in normal cells, elucidating why certain cancers lose maspin may prove to be a more challenging task. Invasive cancer cells undergo numerous genetic changes such as mutation and chromosome rearrangement; the mechanism for loss of maspin expression in these cells is more complicated than just loss of a single transcription factor such as ETS. Indeed, one recent study has shown that in certain breast cancer cells, methylation of maspin promoter may contribute to the silencing of maspin in breast cancers.[40]

About ten-years ago, Dr. Ruth Sager proposed that tumor suppressor genes could be divided into two categories: class I and class II. Class I genes are mutated or deleted in tumors; examples include Rb and p53, which were identified by virtue of their mutations or deletions. Class II genes are not altered at the DNA level; rather, they affect the phenotype by changes in their expression levels.[41] The focus of expression genetics is to re-express the class II tumor suppressor gene in cancers. Since the maspin gene is not mutated or deleted in tumors, it offers great advantage for one to induce maspin expression in tumors by the reagents identified from the above studies. In the future, some natural products or nontoxic drugs will likely be discovered to accomplish this goal.

Role of Maspin in Normal Development

From a genetic point of view, cancer is a developmental abnormality. To understand maspin's role in tumor progression, one needs to elucidate its normal function in organ development.

The mouse provides a model system for genetic manipulation. In comparison with human maspin, study of maspin in mouse has obvious advantages: 1) hormonal regulation of maspin is very difficult to study in human, while it is easy to obtain samples from mouse tissues such as mammary glands at different developmental stages; 2) transgenic animal can be produced to study the effect of overexpression of maspin on mammary gland development and tumor inhibition. Furthermore, knock out mice can be produced to study the loss of function of maspin. For this purpose we cloned the mouse homolog of maspin.

Mouse Maspin is Highly Homologous to Human Maspin

Mouse maspin (mMaspin) cDNA was isolated from a mouse mammary gland cDNA library using human maspin cDNA as a probe. The longest cDNA sequence contained 1378 bp, including 69 bp 5' untranslated region, 1128 bp in the coding region, and 181 bp in the 3'-untranslated region. The poly (A) was added 23 bp after the polyadenylation signal site. The deduced amino acid sequence of mMaspin has a 89% identity with its human counterpart. In addition, a major difference in the mRNA lies in the 3' untranslated region. mMaspin cDNA has a short 3'-UTR, compared to the 1.2 kb 3'-UTR in the human maspin cDNA.[16]

Tissue Distribution

The maspin tissue expression pattern was examined by Northern blot analysis using RNAs from both human and mouse tissues. The human RNA blots from Clontech, Inc. contained a relatively small collection of tissues. Maspin was found to be abundantly expressed in prostate, thymus, testis, and small intestine.[17] Among mouse tissues, the mammary gland has the highest level of expression. Maspin level varied during mammary gland development, suggesting its expression might be hormonally regulated. mMaspin is also expressed in mouse large intestine, skin, tongue, and stomach, and weakly in other tissues such as lung and thymus.[16] It should be pointed out that Northern blot is not the ideal method to characterize maspin expression pattern. We have found out that some tissues express maspin at such a low level that it can not be detected by Northern blot analysis. One example is ovary, which has very low levels of maspin. However, deletion of one copy of maspin in heterozygous mice is enough to cause a severe defect in ovarian development (Zhang et al, unpublished data). Another feature noticed from Northern blot analysis was that there were multiple maspin transcripts in both human and mouse tissues.[16,17] Alternative splicing may account for the multiplicity. Whether the splicing products play any biological role remains to be identified.

Functional Similarity Between Mouse and Human Maspin

Does mouse maspin function similarly as human maspin? To answer this question, we first compared the changes of maspin expression during mouse and human breast cancer progression.

Northern blot analysis showed that mouse maspin was expressed at a high level in mouse normal mammary epithelial cells (NMuMG, HC11). The level of expression was down-regulated in low invasive mouse mammary tumor cells (CSML0) and lost in a highly invasive tumor cell line (CSML100). The down-regulation of maspin protein was also confirmed by Western blot analysis. These data indicate that mouse maspin, like its human homolog is also down-regulated during tumor progression.

Recombinant mMaspin proteins were produced in *E. coli.* as a GST-maspin fusion protein. The biological function of recombinant protein was investigated using Boyden chamber assay (Fig. 3). A pair of mouse mammary tumor cells, CSML0 and CSML100 which are well characterized for their invasiveness, were used in both invasion and migration assays.[42] Normally, about 10% of the low invasive CSML0 cells and 32% of the high invasive CSML100 cells seeded in the upper chamber invade the reconstituted basement membrane. Addition of mMaspin to the cell culture inhibits the invasion of both CSML0 and CSML100 cells. The inhibition by both cell lines increased as the concentration of GST-mMaspin was increased. The highest inhibitions at the concentration of 1.2 μM were about 60% for CSML0 cells (from 10% to 4%) and 62% for CSML100 cells (from 62% to 12.5%). For the motility assay, the recombinant proteins were added at different dosages to CSML0 and CSML100 cell suspensions and incubated with cells for 30 mins at room temperature before seeding. Under control condition, 5.8 % CSML0 cells and 7.2% CSML100 cells move through the polycarbonate membrane. Addition of GST-mMaspin inhibits the motility of both CSML0 and CSML100 cells. The highest inhibitions at the concentration of 1.2 μM were about 78% for CSML0 cells (from 5.9% to 1.25%) and 80% for CSML100 cells (from 7.2% to 1.4%).

These data strongly suggest that mouse maspin functions like human maspin. Therefore, studying the function of maspin with transgenic and knockout mice should provide valid insights into its role in the normal human breast and in breast cancers.

Transgenic Maspin Mice and the Role of Maspin in Normal Mammary Gland Development

Why Study the Role of Maspin in Normal Mammary Gland Development?

The mouse mammary gland undergoes a dramatic series of cyclical changes during development (see Fig. 4). In juvenile mice, the mammary gland develops based on a rapid, hormonally-regulated growth and morphogenesis of the epithelial ducts. Mammary ducts penetrate the gland by means of end buds. Upon reaching sexual maturity, the epithelial ducts have filled the whole gland with a tree-like structure. The onset of pregnancy initiates another stage of mammary development in which

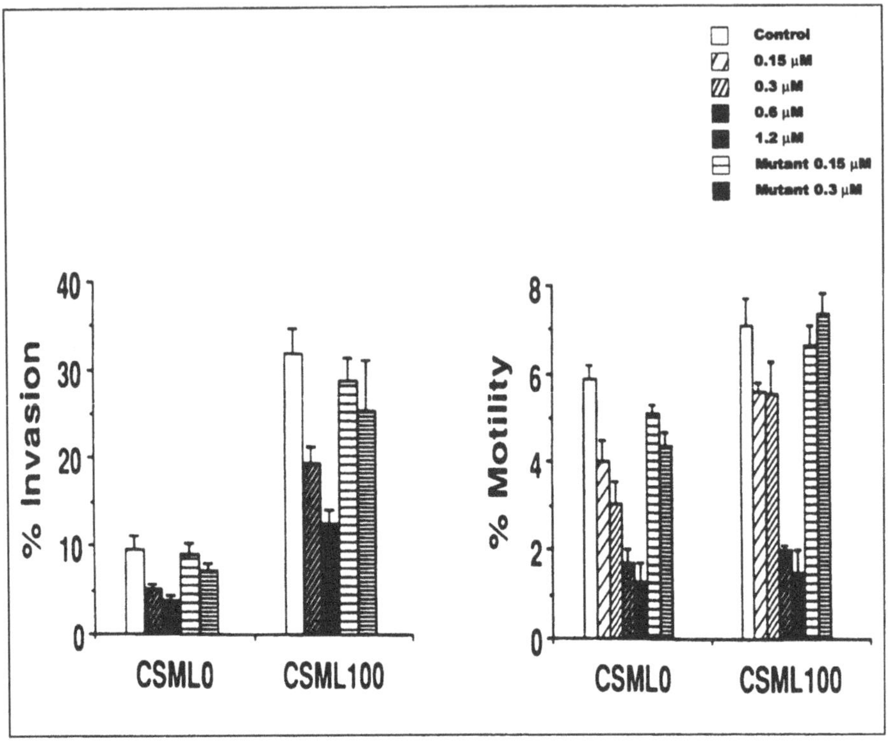

Fig. 3. Effect of mouse mMaspin protein in motility and invasion by two mouse mammary tumor cell lines, CSML0 and CSML100. Control (untreated cells)and maspin treated cells at various concentrations were used in the assay. Mutant maspin is GST-mMaspinΔRSL. Each value represents the averages from two independent assays. In each assay, protein samples were added in triplicate. Error bars stand for standard error of the means.

alveolar epithelial cells rapidly proliferate and differentiate under different regulatory mechanisms. Alveolar cells grow from the ductal skeleton, and appear as lobulo-alveolar units by the end of pregnancy. These alveoli are the functional unit of milk production at lactation. Following lactation, the mammary gland undergoes massive remodeling and apoptosis, resulting in involution of the gland and a return to the ductal structure similar to the non-pregnant state.

One of the barriers to progress in breast cancer research is lack of the biological study of genes in normal mammary gland development. Since maspin expression is regulated during mammary gland development,[43] it is absolutely relevant to examine the role of maspin in this process in order to better understand its involvement in breast cancer.

Establishing Maspin Transgenic Mice

To delineate maspin's function in vivo, we utilized a transgenic mouse system to examine the effect of overexpression of maspin under control of the whey acidic protein (WAP) promoter. The whey acidic protein has been shown to be exclusively expressed in mammary epithelial cells during midpregnancy and lactation.[44,45] Transgenic

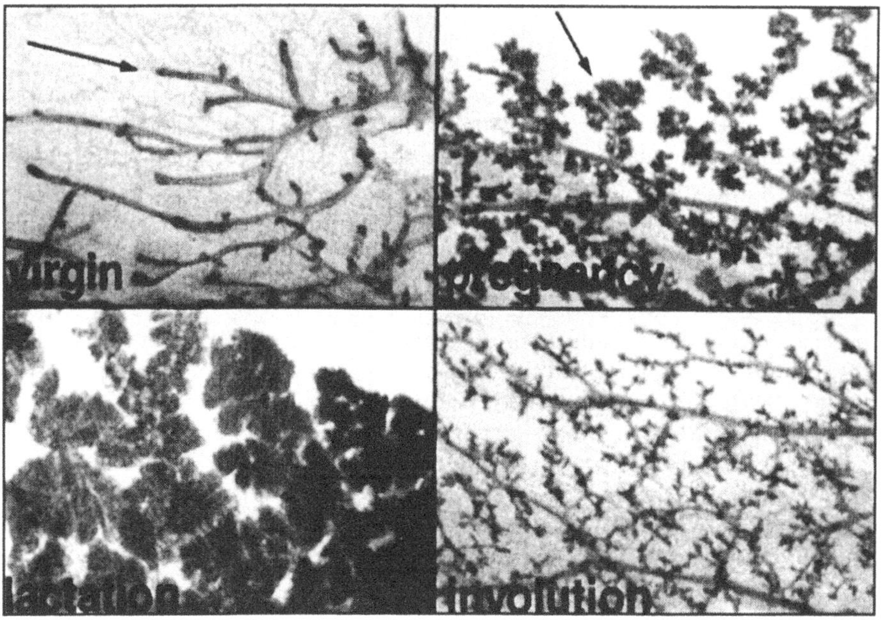

Fig. 4. Mouse normal mammary gland development. Adapted from the website: http://mammary.nih.gov/atlas.

mice were generated by injecting embryos with the WAP-maspin construct. Screening of founders was performed by Southern blot analysis using DNA isolated from mouse tissues. The expression of both endogenous maspin and the transgene in transgenic mouse mammary glands were examined qualitatively by reverse transcription-PCR. RNA was isolated from the mammary glands of transgenic mice at different stages of development. As expected, the WAP-maspin transgene was expressed during late pregnancy and lactation, but was not detectable in virgin mice and glands undergoing involution.

Whole mount and histology of the mammary glands from wild-type and transgenic mice were analyzed. Ductal elongation and branching appeared to be normal in wild-type and transgenic virgin mice. No difference in alveolar structures were present between the transgenic and wild type animals up to day 10 of pregnancy; however, minor differences did become noticeable at day 15 of pregnancy following the activation of the WAP promoter-driven transgene. These mammary glands exhibited decreased alveolar densities, which was further reduced as compared to controls at day 19 and resembled the morphology of the midpregnant wildtype controls. At high power, the mammary glands from transgenic mice contained not only fewer lobular-alveoli structures, but also the size of each alveolar structure was greatly reduced (Fig. 5). In many cases, the lumens of the alveoli were closed. This defect was due to the expression of the transgene activated by WAP promoter from midpregnancy.

Increased Apoptosis and Proliferation in Midpregnant and Early Lactating Mammary Glands of Transgenic Mice

Since the underdevelopment of the mammary glands in the WAP-maspin transgenic mice could have arisen from either decreased proliferation or increased apoptosis or a combination of the two, TUNEL and PCNA immunohistochemistry assays were carried out utilizing pregnant and early lactating mammary glands from wild-type and transgenic mice. As shown in Table 1, the apoptotic rate was significantly increased in transgenic glands at midpregnancy (2.29±0.26%) as compared to controls (0.91± 0.09%). In contrast, little difference was observed in cell proliferation at day 15 of pregnancy. However, during lactation, the apoptosis and proliferation profiles changed significantly in the transgenic strain. Secretive alveolar cells occupied the majority of the fat pad and there was a low rate of proliferation and apoptosis in samples taken from normal mammary glands. This observation contrasted with the results obtained from the WAP-maspin mice, in which a large percentage of the fat pad was devoid of alveolar cells and an increased rate of proliferation was observed. Both proliferation and apoptosis index decreased quickly as lactation proceeded and by lactation day 10, very few cells were PCNA-positive and apoptotic in both transgenic and normal mice.

Effect of Transgene Expression on Milk Gene Expression

The defect in alveolar structures in the WAP-maspin mice during late pregnancy severely hampered the ability of the mother to successfully nurse her entire litter. Indeed, most of the pups died due to insufficient milk production. However, these pups could be rescued by fostering them to a BALB/c non-transgenic lactating female. The number of pups that a transgenic mother could nurse varied between animals. A survey of five sibling mothers at their first pregnancy yielded an average survival rate of 3.6 pups/litter.

Since milk protein genes can function as differentiation markers for the mammary gland, we compared their expression patterns in transgenic and wildtype control mice. Western blot analysis showed that WAP and β-casein were highly expressed in wild-type mammary glands at day 19 of pregnancy and during lactation. However, WAP and β-casein were not detectable in day 19 pregnant transgenic mice. Both milk proteins were present in lactating day 1 transgenic glands, but at a reduced levels, which increased as lactation progressed. This observed decrease was likely due to the effect of reduced number of alveolar cells and closed lumens in the late pregnancy transgenic mice.

Hypothesis for Maspin Function in Mammary Gland

We hypothesize that maspin may primarily regulate cell adhesion and motility in mammary cells, possibly by regulating integrin profiles. Indeed, expression of a dominant negative β1 integrin in the mammary gland, which disrupted the function of β1 and its associated integrins, resulted in a phenotype similar to that observed in the WAP-maspin transgenic mice.[46] Both maspin and chimeric β1 transgene expression caused underdevelopment of the mammary gland in midpregnancy and early lactation, which was accompanied by an increase in apoptosis. In early lactation, milk protein levels were also reduced. The similarity in phenotypes suggests that overexpression of

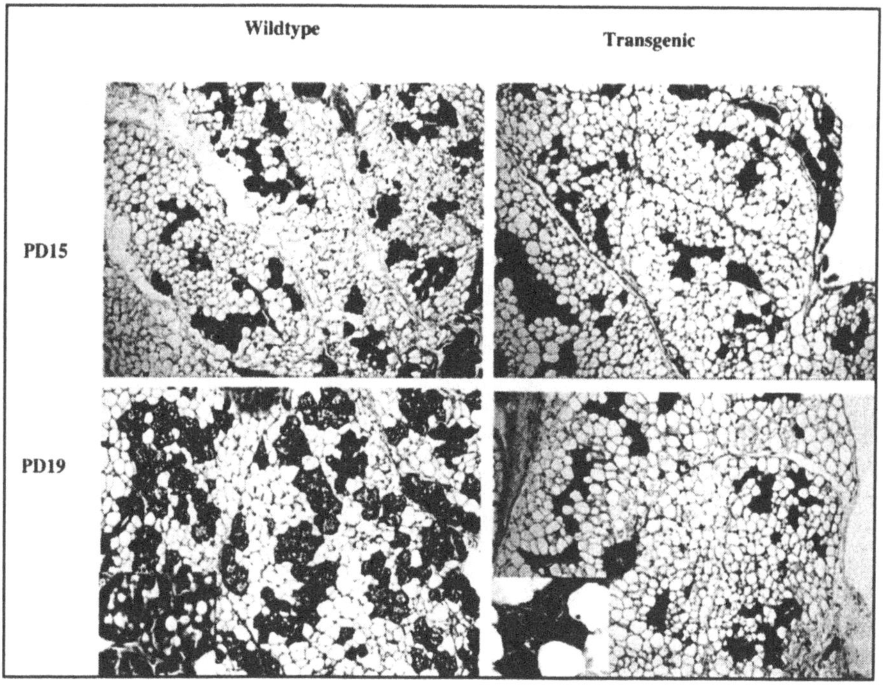

Fig. 5. Histological analysis of mammary tissues from the following mice: (A) wildtype at day 15 pregnancy; (B) transgenics at 15 day pregnancy; (C) wildtype at day 19 pregnancy; (D) transgenics at day 19 pregnancy. (E) and (F) (same magnification) were high power pictures from (C) and (D) respectively. Note the reduced numbers of alveolar structures and the smaller lumen size in the transgenics (D,F). Photographs were taken with a 10X objective for (A-D) and with a 40X objective for (E-F).

maspin may act to perturb integrin regulation or other associated cell adhesion molecules. This hypothesis is partially supported by an in vitro study, which showed that exogenous maspin modified cell attachment to fibronection by regulating integrin profiles, including down-regulation of β1 integrin.[47]

The mechanism by which maspin regulates apoptosis is also unknown. One possibility is that overexpression of maspin perturbs the adhesion of alveolar cells to the ECM as does the chimeric β1 transgene and, thus inhibits the motility of alveolar cells at a stage when invasion into the fat pad is critical. The proliferating alveolar cells are unable to migrate out, leading to increased apoptosis and a resulting small lumen. This is consistent with the concept that proper interactions of mammary epithelial cells with the basement membrane are essential for cell survival, and their disruption will trigger signals leading to apoptosis.[48,49]

Verification of these hypotheses requires identification of new maspin target proteins involved in cell adhesion and migration. We believe maspin possesses multiple functions and may use different domains for these functions (Zhang et al, unpublished data).[18] Understanding the mechanism of maspin action will be greatly facilitated by the analysis of mammary phenotypes using maspin knockout mice.

Table 1. The rate of apoptosis in wildtype and transgenic mice

	WT	Transgenic	P value
15 day pregnant	0.91±0.09 (3)	2.29±0.26 (3)	P<0.01
19 day pregnant	0.54±0.05 (2)	0.91±0.54 (3)	P<0.04
Day 1 lactation	0.15±0.05 (2)	0.79±0.11 (3)	P<0.01

Values are presented as percentage of apoptotic cells (mean ±SD). Number of animals analyzed is indicated in parentheses. About 1000-1200 cells per samples were counted. Statistical analysis was done by student t-test.

Maspin Knockout Mice

A tumor suppressor by definition is a gene when deleted or inactivated, would result in or facilitate the tumorigenesis of an organism. The two-hit hypothesis by Knudson suggests that loss of heterozygosity in tumor suppressor gene will render the organism more susceptible to tumorigenesis. This prediction has been convincingly proved by numerous studies. For example, Donehower et al has demonstrated that p53 heterozygous mice can promote cancer formation and that loss of both p53 alleles is not a prerequisite.[50] To prove maspin does function as a tumor suppressor, we applied gene disruption technique in mice to examine the effect of loss of the maspin function on mammary tumor progression, as well as other mouse development processes. Because of the difficulty to delete a large DNA fragment containing full length of gene, it is a conventional wisdom to target the deletion to a small region of genomic DNA. We have cloned mouse maspin genomic DNAs and designed a vector to selectively knock out the last exon encoding the RSL region and stop codon as well as the entire 3'-UTR. Since 3-UTR is responsible for mRNA stability, it is therefore unlikely that a truncated maspin protein can be made. To this date, we have already generated maspin knockout mice by the approach described above with help from Dr. Philip Leder at Harvard Medical School and Dr. Francesco DeMayo at the Baylor College of Medicine.

The phenotype of the maspin knockout mice is surprising. The homozygous mice are embryonic lethal. The existing knockout mice for serpins and their protease targets are all viable. This demonstrates maspin indeed is a very unique and important serpin. Preliminary data by this laboratory indicate maspin is expressed in early stages of embryonic development (Zhang et al, unpublished data), and lethality demonstrates that maspin plays a critical role. Partial loss of maspin in heterozygotes also displays a phenotype. The heterozygous mothers are less likely to get pregnant and if they do, they deliver small litters of less than 4 pups, which can not be explained by the loss of homozygotes since wildtypes generally produce over 8 pups. Our data indicate that the ovulation efficiency of maspin heterozygotes is greatly reduced compared to control. Histology showed more than 50% of corpus luteum had entrapped oocytes along with hemorrhage in the center of corpus luteum, indicating ovary angiogenesis was affected. Using combined in vitro and in vivo approaches, we find out that maspin indeed plays important role in the process of lutealization and atresia during ovarian development (Zhang et al, unpublished data). We are currently investigating the effect of loss of heterozygosity on mammary gland development and mammary tumor progression.

Role of Maspin in Tumor Progression and Angiogenesis

Mounting evidence has been collected about maspin function in tumor progression.[51] In vitro cell culture studies demonstrated maspin inhibited tumor cell migration and invasion.[15,16,52] Previously, many groups have reported the association of higher maspin level with better tumor prognosis.[37,38,53] The key questions we would like to address are: 1) Does maspin inhibit tumor progression in immunocompetent mice? 2) How does maspin inhibit tumor progression?

Over-Expression of Maspin on Tumor Progression in Bitransgenic Mice

To characterize maspin's in vivo activity during tumor progression, we generated transgenic maspin mice exhibiting targeted overexpression of maspin to mammary epithelial cells under the control of mammary specific whey acidic protein (WAP) promoter.[43] Overexpression of maspin in normal mammary epithelial cells inhibits mammary gland development and induces apoptosis. Since the primary goal of generating WAP-maspin transgenic mice was to test the protective role of maspin overexpression on mammary tumor progression, we crossed WAP-maspin transgenic mice with a strain of oncogenic WAP-simian virus (SV) 40 T antigen (TAg) mice. WAP-TAg transgenic mice develop mammary tumors with 100% frequency and can be utilized to examine specific mechanisms of tumor progression at both early and late time points.[54,55] The SV40 TAg initiates tumorigenesis through the inactivation of both p53 and the pRb related family of proteins.[56,57] In human breast cancers inactivation of p53 function is found in up to 40% of tumors, and mutations in Rb or related proteins have also been reported.[57,58]

Over-Expression of Maspin Reduced Tumor Growth and Vessel Density, and Increased Apoptosis in Primary Adenocarcinomas

Both maspin overexpression and dominant gain of TAg expression induce mammary apoptosis, likely through different mechanisms. To determine if maspin overexpression could increase the level of TAg induced apoptosis in mammary epithelial cells, the percentage of cells undergoing apoptosis in WAP-maspin/WAP-TAg double transgenic mice was compared to WAP-TAg single transgenic mice. Maspin overexpression increased the percentage of apoptotic cells from 1.8% in the single WAP-TAg transgenic mice to 5.6% in the WAP-maspin/WAP-TAg double transgenic mice ($P< 0.01$). In addition, bitransgenic mice exhibited a significant reduction in the number of alveolar-lobular structures and alveolar lumen size when compared to WAP-TAg single transgenic mice. Like WAP-TAg single transgenic mice, none of the WAP-maspin/WAP-TAg double transgenic mice could nurse their offspring beginning with the first litters. These results demonstrate that overexpression of maspin can increase the rate of apoptosis in preneoplastic mammary gland.

We then compared the rate of tumor growth of WAP-TAg and bitransgenic mice (Table 2). Tumors were measured biweekly after they were initially detected and mice were euthanized when the primary tumor reached 2.5 cm in diameter. The tumor observation time (from the time of first palpable tumor to the time when the tumor reached 2.5 cm in diameter) was used to assess the rate of tumor growth. The presence

Table 2. Analysis of primary and lung tumors

	Tag	Bitransgenic	Difference
Tumor-free period (days)	141.3±8.5 (N=15)	140.7±12.7 (N=27)	0.42%
Tumor observation period (days) ♦	36.3±4.2 (N=15)	49.3±4.3 (N=27)	35.9%
Incidence of lung metastasis	55.6%, 15/26 (N=26)	37.5%, 15/40	32.6%
Foci/lung area (10^4 pixels) *	0.66±0.21 (N=15)	0.36±0.08 (N=15)	45.7%

♦ Observation period from the time of 1st tumor appearance to the time when animal was sacrificed. P<0.03.
* The image of total lung area was captured from microscope and was quantitated by NIH Image software. P<0.09.

of maspin overexpression significantly increased the time for a first palpable tumor to reach 2.5 cm in diameter from 36.3 to 49.3 days ($p < 0.03$).

To determine if maspin overexpression has an effect on microvessel density and apoptosis during tumor progression in vivo, microvessel density in adenocarcinomas with and without maspin overexpression were compared. Microvessel density was measured after CD31 staining. In small mammary tumors (tumor size £ 0.6 cm in diameter), the microvessel density was significantly reduced ($p < 0.02$), while the apoptotic index was significantly increased in the presence of maspin overexpression ($p < 0.01$).

Maspin Over-Expression Decreased the Extent of Pulmonary Metastases but Did Not Affect Tumor Initiation

None of the TAg-/maspin+ and TAg-/maspin- mice developed any tumors after 1 year observation. In a separate experiment, 25 WAP-maspin transgenic mice (C57BL/6 background) were followed for more than 2 years, none of the mice developed spontaneous mammary tumors. However, both WAP-TAg and bitransgenic mice developed tumors at similar age. Within 24 weeks after they became pregnant, most of the WAP-TAg and bitransgenic mice developed mammary tumors. Thus, tumor initiation is TAg oncogene dependent.

To detect lung tumor metastasis, lung tissues were sectioned for microscopic analysis (Table 2). At the end point (all mice were sacrificed when primary tumor reached about 2.5 cm in diameter), bitransgenic mice did not show any difference from WAP-TAg mice in the number and weight of mammary tumors developed (data not shown). Serial sections were selected to score for micrometastatic tumor foci under high power microscope. Fifteen out of 26 (55.6%) WAP-TAg mice developed lung metastasis while in bitransgenic mice the rate of metastasis was greatly reduced to 37.5% (15 out of 40 bitransgenic mice). To compare the difference in the number of tumor foci of lung samples between these two species, the microscopic images of lung sections were captured and the foci number/lung area was quantitated for each mouse. The bitransgenic mice had decreased foci numbers (0.356 /10^4 pixels) compared to

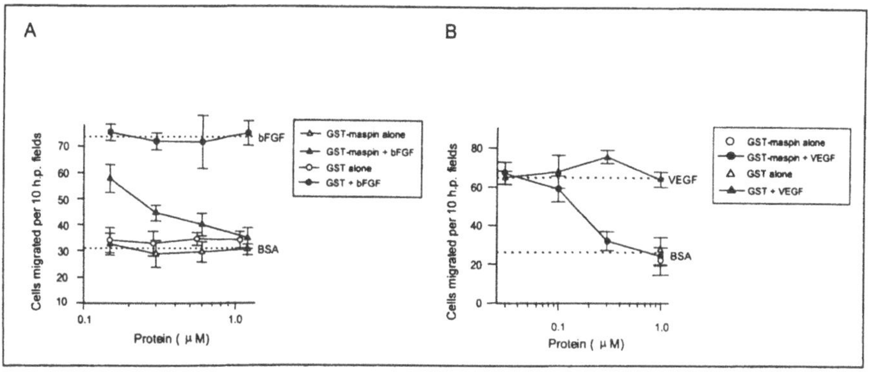

Fig. 6. Effect of maspin on endothelial cell migration. GST-Maspin was tested at a range of concentrations for ability to inhibit endothelial cell migration induced by bFGF (A) or VEGF (B). bFGF, VEGF—indictes migration towards bFGF or VEGF alone; BSA—indicates background migration in the absence of a gradient.

that of WAP-TAg mice (0.655 /10^4 pixels). This information should be useful for future metastasis study.

Maspin as an Angiogenesis Inhibitor

Tumor metastasis is a multi-step process. Malignant tumors need to invade through the basement membrane, stromal tissues, and metastasize to other organs.[59,60] However, for this process to occur, neovascular formation is required.[61,62] This process, termed angiogenesis, provides tumors with nutrients and aids in the removal of metabolic wastes.[63,64] The interaction between endothelial cells and tumor cells also generates a paracrine effect.[65,66]

To study the potential anti-angiogenic properties of maspin, the mouse maspin was produced in *E. coli* as a recombinant GST fusion protein and tested in a variety of angiogenesis assays. Recombinant maspin blocked endothelial cell migration induced by VEGF and bFGF in a dose dependent manner with an ED_{50} of 0.2 μM-0.3 μM (Fig. 6). At 1 μM, maspin completely blocked the response of the endothelial cells to both angiogenic inducers, while the GST control was inactive. In vivo, purified maspin effectively inhibited neovascularization. Rat corneas were surgically implanted with non-inflammatory slow release pellets containing maspin with bFGF and examined six or seven days later for ingrowth of vessels. As shown in Fig. 7, maspin completely blocked bFGF-induced neovascularization. To determine if the anti-angiogenic activity of maspin depended on the inhibition of some undefined protease, several mutants were constructed, expressed and tested. The RSL (reactive serpin loop) near the C-terminus of serpin family members is essential for their anti-protease activity. Mutations at the RSL region of other serpins especially at the P1 site abolish serpin activity.[67] To disrupt this loop in maspin two different mutations were introduced in the RSL region: a C-terminal deletion downstream of P7' residue[16] (maspin DRSL) and a conversion of the critical P1 arginine of the RSL loop to a glutamine (maspin*). A third mutant was constructed in which the first 139 amino acids have been removed but the serpin region left intact (maspinDN). These defective mutants were tested on endothelial cells (Fig. 8). Those with RSL defects retained the ability to inhibit endot-

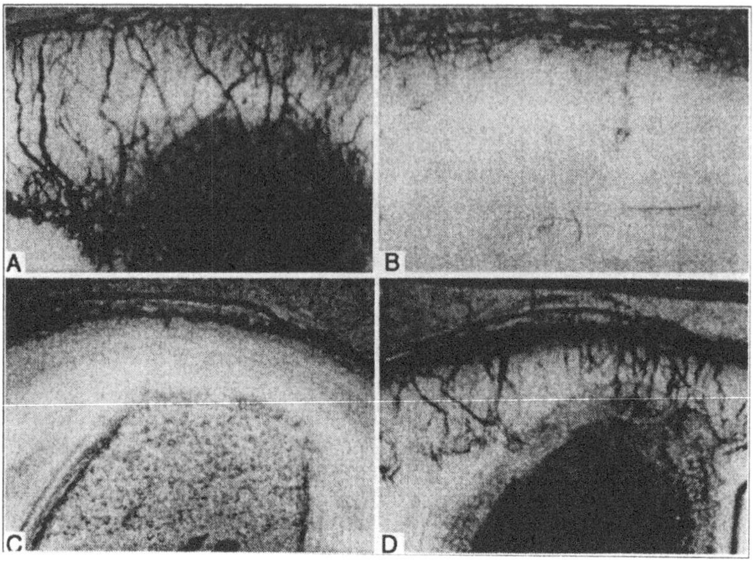

Fig. 7. Maspin inhibition of corneal neovascularization. Pellets containing the indicated test substances at 10μM with bFGF (100 ng/ml) were incorporated into Hydron slow release pellets and implanted in rat corneas. After six or seven days animals were perfused with colloidal carbon to visualize vessels and excised cornease photographed with a 20x objective and scored for neovascularization. A. with bFGF, B. maspin plu bFGF, C. maspinΔN mutant plus bFGF, D. ΔRSL maspin plus bFGF.

helial cell migration and mitogenesis. Protein with mutations in the RSL region also retained the ability to inhibit neovascularization in vivo. The N-terminal deletion, maspinDN, was defective in all assays so it was not possible to determine if a crucial active region had been deleted or if it was not a viable protein.

Complete inhibition of endothelial cell migration in vitro was achieved between 0.5 and 1 μM, in the same concentration range where maspin also inhibits tumor cell motility and invasion,[15] but the mechanisms underlying these two maspin activities seem to be different. We and others find that the former requires that the protein have an intact RSL,[16,52] whereas this feature was not essential for the inhibition of angiogenesis.

To determine if the ability of maspin to inhibit angiogenesis plays a role in its well-documented anti-tumor activity, an athymic mouse xenograft model was utilized. LNCaP prostate tumor cells were implanted subcutaneously on the bidorsal back of nude mice and tumor growth and neovascularization were monitored following systemic treatment with exogenous maspin. We found out that maspin-treated tumors contained significantly fewer vessels as determined by CD31 immunostaining than GST treated controls. To determine whether maspin effects on the tumor-induced vasculature were maintained during a more prolonged treatment, the above experiment was replicated with tumors harvested after 7 to 8 weeks. Thirty-two tumor sites were treated with maspin and 37 with GST. When examined at week 8, the growth of 53% of the maspin-treated tumors had been completely inhibited (Fig. 9). The remaining 15 maspin-treated tumors were reduced in size by average 3.43 fold when compared to GST control treated tumors. The effect of maspin was reversible.

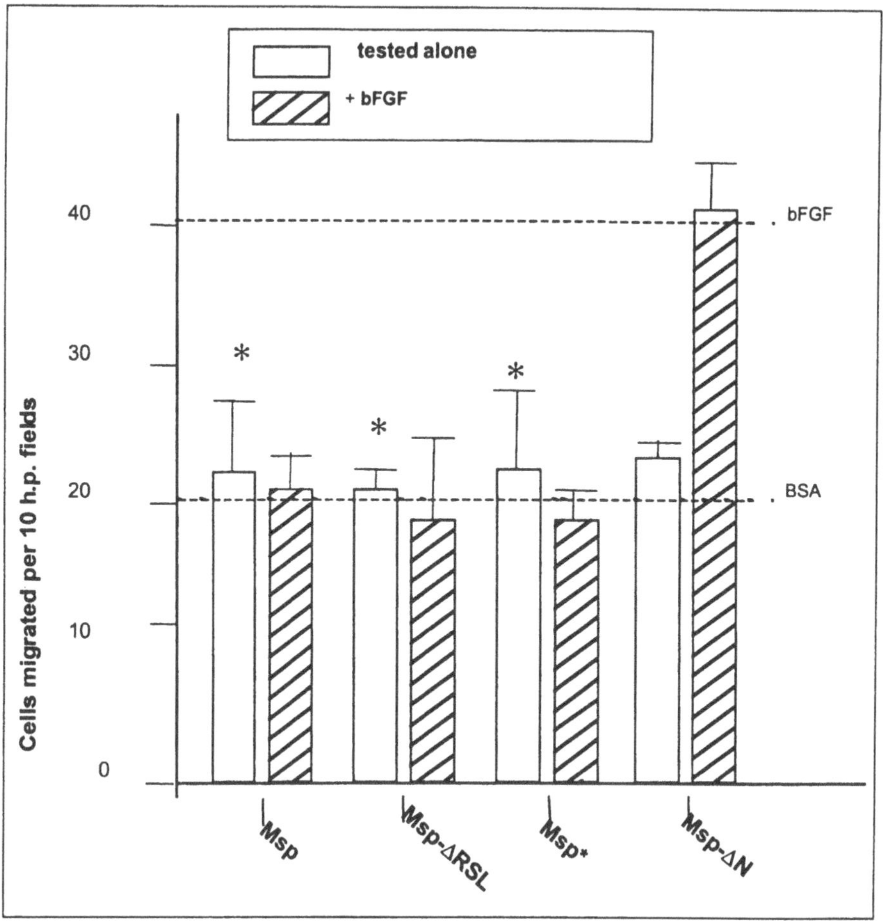

Fig. 8. GST-maspin and its mutants (1 μM) were tested for ability to inhibit endothelial cell migration towards bFGF (10 ng/ml). Maspin, GST-maspin fusion protein; Maspin ΔRSL, maspin with a deletion at C-terminus; Maspin*, maspin containing a R to Q mutation at the P1 residue af the RSL loopl Maspin ΔN, maspin with a deletion at the N-terminus. Glutathalions-S-transferase tested alone was neutral in this assay. * indicates columns significantly different from migration towards bFGF (bFGF----) p<0.01.

To examine if the reduced size of maspin-treated tumors coincided with reduced neovascularization, 20 representative tumors from either maspin-treated (10 sites) or GST-treated tumors (10 sites) were used to quantify the density of microvessels after immunostaining with CD31 antibody. The density of vessels in maspin-treated tumors was reduced 2.6 fold in average to that in control tumors and this difference was highly significant (see Fig. 9). We also compared the treated and control tumors of similar size. A reduction of vessel density was also observed in the maspin treated samples.

The ability to inhibit tumor angiogenesis is only one of several activities associated with the intact maspin protein. Other serpins also have multiple functions and several of them are linked to angiogenesis and tumor growth.[68-71] PAI-1 is involved in modulating both proteolysis and angiogenesis.[12] PEDF, a known regulator of cell dif-

A.

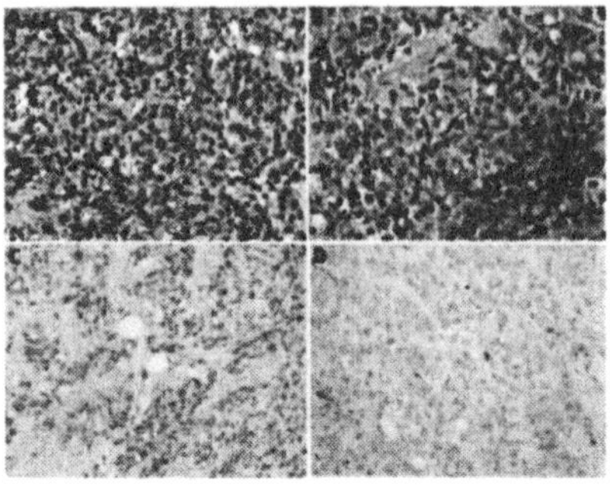

B.

		GST	GST-maspin	P-value
Short Term	Average of MVD*	28.6 ± 3.6 n=10	15.3 ± 1.8 n=10	0.001
Long Term	Total tumor sites	37	32	
	Percentage of complete inhibition**	2.7% (1/37)	53.1% (17/32)	0.0001
	Average of tumor volume (mm^3) of sites with tumors	116.3±26.1, n=36 (12.1-600)	33.9±6.1, n=15 (9.0-87.5)	0.027
	Average of MVD	83.0±10.1, n=10 (38-143)	31.4±1.8, n=10 (21-42)	0.002

* MVD value is collected from three hot fields (400x) in each tumor
** Complete inhibition indicates no sign of tumor or tumor size smaller than 2 mm in diameter after treatment
All ± values are standard error

Fig. 9. Decreased tumor vessels after long-term treatment with exogenous maspin protein. A. Tumors were harvested between 7-8 weeks from GST-treated animals (A,C) and from maspin-treated animals (B,D) and fixed and stained with H & E (A,B) or with anti-CD31 (C,D). B. Analysis of tumor volume and vessel number in GST and GST-maspin treated mice.

ferentiation, is also a very potent anti-angiogenic factor.[72] Such results indicate that a variety of molecules whose structure places them in the serpin family can be important regulators of natural tumor growth via their influence on neovascularization.

Future Directions

Since the discovery of maspin in 1994, a great deal of information has been generated regarding maspin's role in tumorigenesis, metastasis as well as normal development. More than 50 references have been published on maspin study. Despite these advancements, some critical questions remain to be addressed. What is the molecular mechanism of maspin action involved in anti-migration, tumor growth, and angiogenesis? How is maspin gene regulated during mammary gland development? Can we reexpress maspin in cancer cells to revert the phenotype by therapeutic intervention? I hope in the near future my laboratory, along with the colleagues in this field can successfully address these questions through a combined developmental and molecular approach. Understanding them is important for basic biology and will lead to therapies for cancer and other developmental diseases.

References

1. Worrall DM, Blacque OE, Barnes RC. The expanding superfamily of serpins: Searching for the real targets. Biochem Soc Trans 1999; 27(4):746-750.
2. Potempa J, Korzus E, Travis J. The serpin superfamily of proteinase inhibitors: structure, function, and regulation. J Biol Chem 1994; 269(23):15957-15960.
3. Whisstock JC, Irving JA, Bottomley SP, Pike RN, Lesk AM. Serpins in the Caenorhabditis elegans genome. Proteins 1999; 36(1):31-41.
4. Tao W, Walke DW, Morgan JI. Oligomerized Ced-4 kills budding yeast through a caspase-independent mechanism. Biochem Biophys Res Commun 1999; 260(3):799-805.
5. Wang YX, Turner PC, Ness TL, Moon KB, Schoeb TR, Moyer RW. The cowpox virus SPI-3 and myxoma virus SERP1 serpins are not functionally interchangeable despite their similar proteinase inhibition profiles in vitro. Virology 2000; 272(2):281-292.

6. Hunt LT, Dayhoff MO. A surprising new protein superfamily containing ovalbumin, antithrombin- III, and alpha 1-proteinase inhibitor. Biochem Biophys Res Commun 1980; 95(2):864-871.
7. Tombran-Tink J, Li A, Johnson MA, Johnson LV, Chader GJ. Neurotrophic activity of interphotoreceptor matrix on human Y79 retinoblastoma cells. J Comp Neurol 1992; 317(2):175-186.
8. Brunner G, Preissner KT. Pericellular enzymatic hydrolysis: Implications for the regulation of cell proliferation in the vessel wall and the bone marrow. Blood Coagul Fibrinolysis 1994; 5(4):625-639.
9. Blasi F. Urokinase and urokinase receptor: A paracrine/autocrine system regulating cell migration and invasiveness. Bioessays 1993; 15(2):105-111.
10. Deng G, Curriden SA, Wang S, Rosenberg S, Loskutoff DJ. Is plasminogen activator inhibitor-1 the molecular switch that governs urokinase receptor-mediated cell adhesion and release? J Cell Biol 1996; 134(6):1563-1571.
11. Stefansson S, Lawrence DA. The serpin PAI-1 inhibits cell migration by blocking integrin alpha V beta 3 binding to vitronectin [see comments]. Nature 1996; 383(6599):441-443.
12. Bajou K, Noel A, Gerard RD, Masson V, Brunner N, Holst-Hansen C et al. Absence of host plasminogen activator inhibitor 1 prevents cancer invasion and vascularization. Nat Med 1998; 4(8):923-928.
13. Zou Z, Anisowicz A, Hendrix MJ, Thor A, Neveu M, Sheng S et al. Maspin, a serpin with tumor-suppressing activity in human mammary epithelial cells [see comments]. Science 1994; 263(5146):526-529.
14. Sun J, Rose JB, Bird P. Gene structure, chromosomal localization, and expression of the murine homologue of human proteinase inhibitor 6 (PI-6) suggests divergence of PI-6 from the ovalbumin serpins. J Biol Chem 1995; 270(27):16089-16096.
15. Sheng S, Carey J, Seftor EA, Dias L, Hendrix MJ, Sager R. Maspin acts at the cell membrane to inhibit invasion and motility of mammary and prostatic cancer cells. Proc Natl Acad Sci USA 1996; 93(21):11669-11674.
16. Zhang M, Sheng S, Maass N, Sager R. mMaspin: The mouse homolog of a human tumor suppressor gene inhibits mammary tumor invasion and motility. Mol Med 1997; 3(1):49-59.
17. Zhang M, Maass N, Magit D, Sager R. Transactivation through ets and Ap1 transcription sites determines the expression of the tumor-suppressing gene maspin. Cell growth Differ 1997; 8(2):179-186.
18. Zhang M, Volpert O, Shi YH, Bouck N. Maspin is an angiogenesis inhibitor. Nat Med 2000; 6(2):196-199.
19. Sternlicht MD, Kedeshian P, Shao ZM, Safarians S, Barsky SH. The human myoepithelial cell is a natural tumor suppressor. Clin Cancer Res 1997; 3(11):1949-1958.
20. Sheng S, Truong B, Fredrickson D, Wu R, Pardee AB, Sager R. Tissue-type plasminogen activator is a target of the tumor suppressor gene maspin. Proc Natl Acad Sci USA 1998; 95(2):499-504.
21. Pemberton PA, Wong DT, Gibson HL, Kiefer MC, Fitzpatrick PA, Sager R et al. The tumor suppressor maspin does not undergo the stressed to relaxed transition or inhibit trypsin-like serine proteases. Evidence that maspin is not a protease inhibitory serpin. J Biol Chem 1995; 270(26):15832-15837.
22. Hopkins PC, Whisstock J. Function of maspin [letter; comment]. Science 1994; 265(5180):1893-1894.
23. Bartuski AJ, Kamachi Y, Schick C, Overhauser J, Silverman GA. Cytoplasmic antiproteinase 2 (PI8) and bomapin (PI10) map to the serpin cluster at 18q21.3. Genomics 1997; 43(3):321-328.
24. Abts HF, Welss T, Mirmohammadsadegh A, Kohrer K, Michel G, Ruzicka T. Cloning and characterization of hurpin (protease inhibitor 13): A new skin-specific, UV-repressible serine proteinase inhibitor of the ovalbumin serpin family. J Mol Biol 1999; 293(1):29-39.
25. Scott FL, Eyre HJ, Lioumi M, Ragoussis J, Irving JA, Sutherland GA et al. Human ovalbumin serpin evolution: Phylogenic analysis, gene organization, and identification of new PI8-related genes suggest that two interchromosomal and several intrachromosomal duplications generated the gene clusters at 18q21-q23 and 6p25. Genomics 1999; 62(3):490-499.

26. Schleef RR, Chuang TL. Protease inhibitor 10 inhibits tumor necrosis factor alpha-induced cell death. Evidence for the formation of intracellular high M(r) protease inhibitor 10-containing complexes. J Biol Chem 2000; 275(34):26385-26389.
27. Sprecher CA, Morgenstern KA, Mathewes S, Dahlen JR, Schrader SK, Foster DC et al. Molecular cloning, expression, and partial characterization of two novel members of the ovalbumin family of serine proteinase inhibitors. J Biol Chem 1995; 270(50):29854-29861.
28. Harrop SJ, Jankova L, Coles M, Jardine D, Whittaker JS, Gould AR et al. The crystal structure of plasminogen activator inhibitor 2 at 2.0 A resolution: implications for serpin function. Structure Fold Des 1999; 7(1):43-54.
29. Sun J, Ooms L, Bird CH, Sutton VR, Trapani JA, Bird PI. A new family of 10 murine ovalbumin serpins includes two homologs of proteinase inhibitor 8 and two homologs of the granzyme B inhibitor (proteinase inhibitor 9). J Biol Chem 1997; 272(24):15434-15441.
30. Schulze H, Isaacs JT, Coffey DS. A critical review of the concept of total androgen ablation in the treatment of prostate cancer. Prog Clin Biol Res 1987; 9:1-19.
31. Beato M. Gene regulation by steroid hormones. Cell 1989; 56(3):335-344.
32. Laborda J. 36B4 cDNA used as an estradiol-independent mRNA control is the cDNA for human acidic ribosomal phosphoprotein PO. Nucleic Acids Res 1991; 19(14):3998.
33. Drouin J, Sun YL, Chamberland M, Gauthier Y, De Lean A, Nemer M et al. Novel glucocorticoid receptor complex with DNA element of the hormone-repressed POMC gene. Embo J 1993; 12(1):145-156.
34. Starr DB, Matsui W, Thomas JR, Yamamoto KR. Intracellular receptors use a common mechanism to interpret signaling information at response elements. Genes Dev 1996; 10(10):1271-1283.
35. Zhang M, Magit D, Sager R. Expression of maspin in prostate cells is regulated by a positive ets element and a negative hormonal responsive element site recognized by androgen receptor. Proc Natl Acad Sci USA 1997; 94(11):5673-5678.
36. Zou Z, Gao C, Nagaich AK, Connell T, Saito S, Moul JW et al. p53 regulates the expression of the tumor suppressor gene maspin. J Biol Chem 2000; 275(9):6051-6054.
37. Mueller E, Sarraf P, Tontonoz P, Evans RM, Martin KJ, Zhang M et al. Terminal differentiation of human breast cancer through PPAR gamma. Mol Cell 1998; 1(3):465-470.
38. Jiang WG, Hiscox S, Horrobin DF, Bryce RP, Mansel RE. Gamma linolenic acid regulates expression of maspin and the motility of cancer cells. Biochem Biophys Res Commun 1997; 237(3):639-644.
39. Li JJ, Colburn NH, Oberley LW. Maspin gene expression in tumor suppression induced by overexpressing manganese-containing superoxide dismutase cDNA in human breast cancer cells. Carcinogenesis 1998; 19(5):833-839.
40. Domann FE, Rice JC, Hendrix MJ, Futscher BW. Epigenetic silencing of maspin gene expression in human breast cancers. Int J Cancer 2000; 85(6):805-810.
41. Sager R. Tumor suppressor genes: The puzzle and the promise. Science 1989; 246(4936):1406-1412.
42. Tulchinsky E, Ford HL, Kramerov D, Reshetnyak E, Grigorian M, Zain S et al. Transcriptional analysis of the mts1 gene with specific reference to 5' flanking sequences. Proc Natl Acad Sci USA 1992; 89(19):9146-9150.
43. Zhang M, Magit D, Botteri F, Shi Y, He K, Li M et al. Maspin plays an important role in mammary gland development. Developmental Biology 1999; 215:278-287.
44. Pittius CW, Sankaran L, Topper YJ, Hennighausen L. Comparison of the regulation of the whey acidic protein gene with that of a hybrid gene containing the whey acidic protein gene promoter in transgenic mice. Mol Endocrinol 1988; 2(11):1027-1032.
45. Pittius CW, Hennighausen L, Lee E, Westphal H, Nicols E, Vitale J et al. A milk protein gene promoter directs the expression of human tissue plasminogen activator cDNA to the mammary gland in transgenic mice. Proc Natl Acad Sci USA 1988; 85(16):5874-5878.

46. Faraldo MM, Deugnier MA, Lukashev M, Thiery JP, Glukhova MA. Perturbation of beta1-integrin function alters the development of murine mammary gland. Embo J 1998; 17(8):2139-2147.
47. Seftor RE, Seftor EA, Sheng S, Pemberton PA, Sager R, Hendrix MJ. maspin suppresses the invasive phenotype of human breast carcinoma. Cancer Res 1998; 58(24):5681-5685.
48. Frisch SM, Francis H. Disruption of epithelial cell-matrix interactions induces apoptosis. J Cell Biol 1994; 124(4):619-626.
49. Frisch SM, Ruoslahti E. Integrins and anoikis. Curr Opin Cell Biol 1997; 9(5):701-706.
50. Venkatachalam S, Shi YP, Jones SN, Vogel H, Bradley A, Pinkel D et al. Retention of wild-type p53 in tumors from p53 heterozygous mice: reduction of p53 dosage can promote cancer formation. Embo J 1998; 17(16):4657-4667.
51. Hendrix MJ. De-mystifying the mechanism(s) of maspin [news]. Nat Med 2000; 6(4):374-376.
52. Sheng S, Pemberton PA, Sager R. Production, purification, and characterization of recombinant maspin proteins. J Biol Chem 1994; 269(49):30988-30993.
53. Xia W, Lau YK, Hu MC, Li L, Johnston DA, Sheng S et al. High tumoral maspin expression is associated with improved survival of patients with oral squamous cell carcinoma. Oncogene 2000; 19(20):2398-2403.
54. Tzeng YJ, Guhl E, Graessmann M, Graessmann A. Breast cancer formation in transgenic animals induced by the whey acidic protein SV40 T antigen (WAP-SV-T) hybrid gene. Oncogene 1993; 8(7):1965-1971.
55. Li M, Hu J, Heermeier K, Hennighausen L, Furth PA. Expression of a viral oncoprotein during mammary gland development alters cell fate and function: Induction of p53-independent apoptosis is followed by impaired milk protein production in surviving cells. Cell growth Differ 1996; 7(1):3-11.
56. Dyson N, Buchkovich K, Whyte P, Harlow E. The cellular 107K protein that binds to adenovirus E1A also associates with the large T antigens of SV40 and JC virus. Cell 1989; 58(2):249-255.
57. Li M, Lewis B, Capuco AV, Laucirica R, Furth PA. WAP-TAg transgenic mice and the study of dysregulated cell survival, proliferation, and mutation during breast carcinogenesis. Oncogene 2000; 19(8):1010-1019.
58. Lee EY, To H, Shew JY, Bookstein R, Scully P, Lee WH. Inactivation of the retinoblastoma susceptibility gene in human breast cancers. Science 1988; 241(4862):218-221.
59. Stetler-Stevenson WG, Aznavoorian, S., Liotta, L. A. Tumor cell interactions with the extracellular matrix during invasion and metastasis. Ann Rev Cell Biol 1993; 9:541-573.
60. Mignatti P, Rifkin DB. Biology and biochemistry of proteinases in tumor invasion. Physiol Rev 1993; 73(1):161-195.
61. Folkman J. How is blood vessel growth regulated in normal and neoplastic tissue? G.H.A. Clowes memorial Award lecture. Cancer Res 1986; 46(2):467-473.
62. Folkman J. Successful treatment of an angiogenic disease [editorial]. N Engl J Med 1989; 320(18):1211-1212.
63. Folkman J. Tumor angiogenesis. Mol Basis Cancer 1995:206-232.
64. Fidler IJ, Ellis LM. The implications of angiogenesis for the biology and therapy of cancer metastasis [comment]. Cell 1994; 79(2):185-188.
65. Jones A, Harris AL. New developments in angiogenesis: a major mechanism for tumor growth and target for therapy. Cancer J Sci Am 1998; 4(4):209-217.
66. Ambs S, Bennett WP, Merriam WG, Ogunfusika MO, Oser SM, Khan MA et al. Vascular endothelial growth factor and nitric oxide synthase expression in human lung cancer and the relation to p53. Br J Cancer 1998; 78(2):233-239.
67. Stringer HA, Pannekoek H. The significance of fibrin binding by plasminogen activator inhibitor 1 for the mechanism of tissue-type plasminogen activator-mediated fibrinolysis. J Biol Chem 1995; 270(19):11205-11208.

68. Carmeliet P, Schoonjans L, Kieckens L, Ream B, Degen J, Bronson R et al. Physiological consequences of loss of plasminogen activator gene function in mice. Nature 1994; 368(6470):419-424.
69. Koolwijk P, van Erck MG, de Vree WJ, Vermeer MA, Weich HA, Hanemaaijer R et al. Cooperative effect of TNFalpha, bFGF, and VEGF on the formation of tubular structures of human microvascular endothelial cells in a fibrin matrix. Role of urokinase activity. J Cell Biol 1996; 132(6):1177-1188.
70. Montesano R, Pepper MS, Mohle-Steinlein U, Risau W, Wagner EF, Orci L. Increased proteolytic activity is responsible for the aberrant morphogenetic behavior of endothelial cells expressing the middle T oncogene. Cell 1990; 62(3):435-445.
71. Min HY, Doyle LV, Vitt CR, Zandonella CL, Stratton-Thomas JR, Shuman MA et al. Urokinase receptor antagonists inhibit angiogenesis and primary tumor growth in syngeneic mice. Cancer Res 1996; 56(10):2428-2433.
72. Dawson DW, Volpert OV, Gillis P, Crawford SE, Xu H, Benedict W et al. Pigment Epithelium-derived factor: A potent inhibitor of angiogenesis. Science 1999; 285(5425):245-248.

Chapter 9

The Role of Maspin in Human Placental Development

Anuja Dokras, Lynn M.G. Gardner, Dawn A. Kirschmann, Elisabeth A. Seftor and Mary J.C. Hendrix

Introduction

The role of maspin in the human reproductive tract is unknown. The human placenta is a physiological site of invasion where cytotrophoblasts invade into maternal tissues during gestation. This process is spatially and temporally regulated but the role of tumor suppressor genes in this invasive model has not been studied in detail. We have examined the expression of maspin in human placenta during all three trimesters of pregnancy and hypothesize a potential role for maspin in regulating invasion at this unique site.

The human placenta is hemochorial and displays highly regulated invasive activity and exponential growth potential. The stem cell cytotrophoblasts undergo differentiation along two pathways: they fuse to form multinucleate syncytiotrophoblasts or they detach from the basement membrane to form mononuclear cell columns and anchoring villi that invade into the endometrium, myometrium, and spiral arteries. Thus, cytotrophoblast invasion physically anchors the fetus to the mother. The depth of placental invasion is precisely controlled and temporally regulated, and abnormalities in invasion can have clinically relevant consequences. For example, shallow invasion of cytotrophoblasts has been reported in conditions such as preeclampsia and intrauterine growth retardation.[1,2] Increased invasion is associated with conditions such as placenta accreta and placenta increta, which are both associated with postPArtum hemorrhage. Despite the extensive spatial invasion, the risk of developing an invasive mole or choriocarcinoma remains low (1 in 20,000-40,000 pregnancies).

Regulation of Cytotrophoblast Invasion

The dynamic process of invasion demonstrated by extravillous cytotrophoblasts appears to be regulated by similar mechanisms as have been described in cancer cells: extracellular matrix (ECM) degrading enzymes, matrix metalloproteases (MMPs), adhesion molecules (cadherins), ECM receptors (integrins), and cytokines. The invasive capacity of cytotrophoblasts, as assessed in vitro, decreases from the first to the third

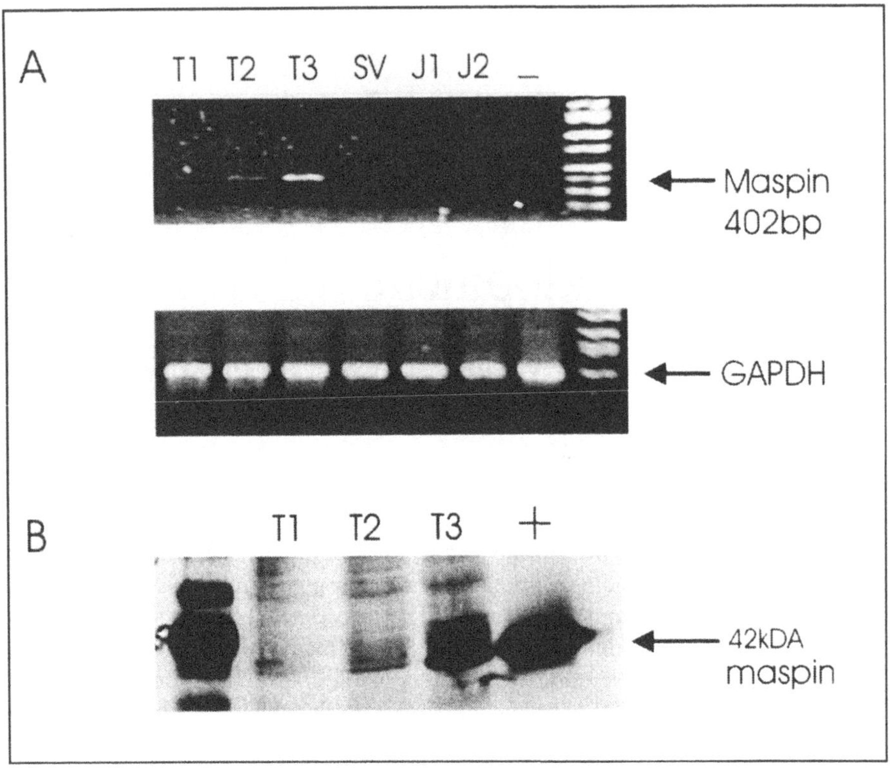

Fig. 1. (A) Maspin mRNA expression in human placenta. Total RNA from human placentae from first (T1), second (T2) and third (T3) trimesters, HTR-8/SVneo (SV), JEG (J1) and JAR (J2) and MDA-MB-231 (invasive breast cancer cell line used as a negative control) were subjected to semiquantitative RT-PCR analysis using maspin-specific primers. GAPDH-specific primers were used to determine equal loading of samples. (B) The expression of maspin protein was measured in equal amounts of lysates obtained from cytotrophoblasts isolated from first (T1), second (T2) and third (T3) trimesters by Western blot analysis. The mouse anti-human maspin monoclonal antibody identified a 42kD band corresponding to the rMaspin used as the positive control (+).

trimester mimicking the in vivo situation.[3] The stem cell cytotrophoblast populations express high levels of E-cadherin and $\alpha_6\beta_4$ integrins.[4] During the process of invasion into the maternal surface, the extravillous cytotrophoblasts down-regulate their expression of E-cadherin and upregulate $\alpha_1\beta_1$ integrins (laminin receptor). Further, invading cytotrophoblasts upregulate the expression of MMPs, especially MMP-9.[5] More importantly, during the first trimester both inhibitors of MMPs, TIMP and TIMP-2, have been shown to decrease cytotrophoblast invasion in vitro. MMPs and their inhibitors are also regulated in the first trimester placenta by the cytokines IL-1β,[6] TGFβ1 and TGFβ2[7] and EGF.[3]

Tumor Suppressor Genes in the Human Placenta

A few tumor suppressor genes have been detected in the human placenta including p53,[8] WT1[9] and maspin.[10] In the placenta, maspin is localized to cytotrophoblasts in chorionic villi.[10] However, there is no information in the literature regarding the

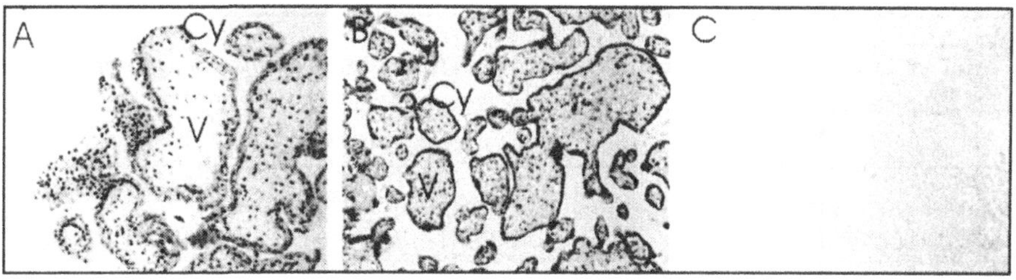

Fig. 2. Immunohistochemistry of human placentae with mouse anti-human maspin monoclonal antibody. (A) First trimester placenta shows faint positive staining for the cytotrophoblast layer (Cy) and occasional cells in the mesenchymal villous core (V). (B) Third trimester villi show increased staining in the syncytio-cytotrophoblast layers. (C) Negative control for the third trimester. Magnification: 200x.

putative biological role of any of the above tumor suppressor genes in the human placenta or the female reproductive tract. Maspin (mammary serine protease inhibitor) is a tumor suppressor gene which has been shown to have inhibitory actions on motility, invasion, metastasis, and angiogenesis in human breast and prostate cancer cells.[11,12] Loss of maspin expression in these cancers correlates with tumor invasiveness and breast cancer recurrence. Further, treatment of human breast and prostate cancer cells with recombinant maspin (rMaspin) inhibited cell motility.[13] These observations suggest that maspin expression plays an important role in regulating tumor cell invasion and metastasis.

Maspin in the Human Placenta

We were interested in examining the role of maspin in regulating cytotrophoblast invasion and therefore determined the expression of maspin in human placenta obtained after first and second trimester terminations (7-20 weeks) and after delivery by cesarean section at term (37-40 weeks). Maspin mRNA expression was maximally detected in third trimester placentae by semi-quantitative RT-PCR, as compared to first and second trimester placental tissues (Fig. 1A). Similar results were obtained when cytotrophoblasts isolated by sequential enzymatic digestion (first and second trimester isolation,[14] third trimester[15]) were examined. In contrast, expression of maspin was not detected in the invasive extravillous cytotrophoblast cell line derived from a first trimester placenta, (HTR-8/SVneo, kind gift from Dr Charles Graham, Canada) nor in the choriocarcinoma cell lines (JEG, JAR) (Fig. 1A).

Consistent with maspin mRNA expression, maspin protein was also maximally detected by Western blot in cell lysates from cytotrophoblasts isolated from term placentae (Fig. 1B). The level of maspin protein was 2-fold higher in the second trimester compared to the first trimester, and 4.5-fold higher in the third trimester compared to the first trimester, as measured by densitometric analysis. In addition, immunohistochemistry was used to localize maspin in placental tissues using a monoclonal mouse anti-human maspin antibody (Pharmingen, San Diego, CA). In the first trimester, maspin staining was patchy and restricted to the cytotrophoblast layer with no staining detected in the syncytiotrophoblast layer (Fig. 2A). There was more uniform staining in the syncytiocytotrophoblast layers later in gestation with maximum staining in the third trimester (Fig. 2B).

Table 1. Effects of recombinant maspin on cytotrophoblast invasion in vitro

	First trimester	Second trimester	Third trimester
Controls	1143±87	1173±123	775±109
Maspin 40 μg/ml	644 ±82	668±66	596±69
Maspin 60 μg/ml	574±40*	581±126**	463±10¶

Values represent Mean ± SD.
*p<0.005,**p<0.01,¶p<0.03

The invasive ability of cytotrophoblasts from all three trimesters was assessed using an in vitro invasion assay, which measured the ability of the cells to traverse a defined human basement membrane matrix over 60 hours using membrane invasion culture system (MICS) chambers.[16] Cytotrophoblasts isolated from the first (mean±SE: 695±37/filter) and second (677±50/filter) trimesters demonstrated a significantly higher invasive ability as compared to those isolated from term placentae (298±77, p<0.02). The invasive ability of both HTR-8/SVneo (16264±1117, p<0.004) and the choriocarcinoma cell line, JEG (2943±634, p<0.03) was significantly higher when compared to first trimester cytotrophoblasts.

Next, we examined the direct effects of rMaspin (Arriva Pharmaceuticals Inc, Alameda, CA) on cytotrophoblast invasion using the same in vitro invasion assay system. There was a dose-dependent decrease in the invasive ability of cytotrophoblasts in all three trimesters receiving rMaspin treatment for 24 hours (Table 1). Upon the addition of 60 μg/ml rMaspin to the test wells, the first and second trimester cytotrophoblasts showed a 50% decrease and the third trimester cytotrophoblasts showed a 40% decrease in invasive ability when compared to the untreated controls.

Discussion

Our data demonstrates for the first time that maspin, a tumor suppressor gene, is differentially expressed in the human placenta and maximally expressed in the third trimester. Furthermore, the in vitro invasion experiments demonstrate that cytotrophoblasts isolated from the third trimester have the least invasive ability when compared to cytotrophoblasts isolated from first and second trimesters. These results indicate that the decreased expression of maspin in the first trimester corresponds to the period of maximum cytotrophoblast invasion as shown in vitro. This hypothesis is further supported by the absence of maspin expression in the invasive cell lines, HTR-8/SVneo and JEG. In addition, we detected a 40-50% decrease in invasive ability with the addition of rMaspin to cytotrophoblasts, as compared to controls. These data suggest that maspin may play a role in the regulation of cell invasion at a physiological site, the human placenta.

Studies from our laboratory have previously shown that treatment of highly invasive breast cancer cells with recombinant maspin causes increased surface expression of the α_5 integrin and down-regulation of MMPs, which converts these cells to a more benign epithelial phenotype.[17] Interestingly, first trimester cytotrophoblasts cultured in vitro rapidly change their integrin repertoire to mimic invasive ($\alpha_1\beta_1$) rather than stem cell cytotrophoblasts ($\alpha_6\beta_4$).[4] However, the authors reported that this ability of the cells to differentiate along the invasive pathway was lost in cytotrophoblasts isolated from term placentae. Hence it is possible that increased expression of maspin at term may regulate cytotrophoblast invasion by altering the surface integrin expression of these cells. The hierarchical role of the above described factors including tumor suppressor genes, cytokines, MMPs, and integrins in human placental development in vivo remains to be determined. The down-regulation of maspin expression may be critical at the time of implantation and early placental development, whereas up-regulation of maspin may provide a signal for cytotrophoblasts to decrease invasive activity at the end of gestation.

Ackowledgements

This work was supported by a grant to Anuja Dokras from the Reproductive Scientist Development Program through NIH grant#5K12HD00849 and the ASRM

References

1. Zhou Y, Damsky CH, Chui, K, Roberts JM, Fisher SJ. Preeclampsia is associated with abnormal expression of adhesion molecules by invasive cytotrophoblasts. J Clin Invest 1993; 91:950-960.
2. Gerretsen G, Huisjes HJ, Elema JD. Morphological changes of the spiral arteries in the placental bed in relation to preeclampsia and fetal growth retardation. Br J Ob Gyn 1981; 88(9):876-881.
3. Bass KE, Morrish D, Roth I, Bhardwaj D, Taylor R, Zhou Y et al. Human cytotrophoblast invasion is up-regulated by epidermal growth factor: Evidence that paracrine factors modify this process. Dev Biol 1994;164:550-561.
4. Damskey CH, Librach C, Lim K-H, Fitzgerald ML, McMaster MT, Janatpour M et al. Integrin switching regulates normal trophoblast invasion. Development 1994; 120:3657-3666.
5. Librach CL, Werb Z, Fitzgerald ML, Chiu K, Corwin NM, Esteves RA et al. 92-kD type IV collagenase mediates invasion of human cytotrophoblasts. J Cell Biol 1991; 113:437-449.
6. Librach CL, Feigenbaum SL, Bass KE, Cui T-Y, Verastas N, Sadovski Y et al. Interleukin-1b regulates human cytotrophoblast metalloproteinase activity and invasion in vitro. J Bioligical Chem 1994; 269:17125-17131.
7. Graham CH, Lala PK. Mechanism of control of trophoblast invasion in situ. J Cell Physiol 1991; 147:228-234.
8. Roncalli M, Bulfamante G, Viale G, Springall DR, Alfano R, Comi A et al. C-myc and tumor suppressor gene product expression in developing and term human trophoblast. Placenta 1994; 15:399-409.
9. Feingold M, Zilberstein M, Srivastava RK, Seibel MM, Bar-Ami S, Hambartsoumian E. Expression of Wilms' tumor suppressor gene (WT-1) in term human trophoblast: Regulation by cyclic adenosine 3',5'-monophosphate. J Clin Endocrin & Metabol 1998; 83:2503-2508.
10. Pemberton PA, Tipton AR, Pavloff N, Smith J, Erickson JR, Mouchabeck ZM, Kiefer MC. Maspin is an intracellular serpin that partitions into secretory vesicles and is present at the cell surface. J Histochem Cytochem 1997; 45:1697-1706.
11. Sternlicht MD, Safarians S, Rivera SP, Barsky SH. Characterizations of the extra-cellular matrix and proteinase inhibitor content of human myoepithelial tumors. Laboratory Invest 1996; 74:781-796.
12. Zhang M, Volpert O, Shi YH, Bouck N. Maspin is an angiogenesis inhibitor. Nat Med 2000; 6(2):196-199.
13. Sheng S, Carey J, Seftor EA, Dias L, Hendrix MJ, Sager R. Maspin acts at the cell membrane to inhibit invasion and motility of mammary and prostatic cancer cells. Proc Natl Acad Sci USA 1996; 93:11669-11674.

14. Fisher SJ, Cui TY, Zhang L, Hartman L, Grahl K, Zhang GY et al. Adhesive and degradative properties of human placental cytotrophoblast cells *in vitro*. J Cell Biol 1989; 109:891-902
15. Kliman HJ, Nestler JE, Sermasi E, Sanger JM, Strauss JF III. Purification, characterization and *in vitro* differentiation of cytotrophoblast from human term placentae. Endocrinology 1986; 118:1567-1582.
16. Hendrix MJC, Seftor EA, Seftor REB, Fidler IJ. A simple quantitative assay for studying the invasive potential of high low human metastatic variants. Cancer Lett 1987; 38:137-147.
17. Seftor RE, Seftor EA, Sheng S, Pemberton PA, Sager R, Hendrix MJC. Maspin suppresses the invasive phenotype of human breast carcinoma. Cancer Res 1998; 58:5681-5685.

Chapter 10

Maspin, a Potential Prognostic Marker for Human Cancers

Mickey C-T. Hu, Weiya Xia and Mien-Chie Hung

Introduction

Maspin (*ma*mmary *s*erine *p*rotease *in*hibitor) is a 42 kDa protein that shares significant sequence homology with several members of the serpin (serine protease inhibitor) family, including plasminogen activation inhibitors 1 and 2 (PAI-1 and PAI-2), a1-antitrypsin, and non-inhibitor proteins such as ovalbumin.[1,2] It is expressed in normal human mammary epithelial cells[2] and is associated with secretory vesicles and cellular surface.[3] It has been shown that maspin associates with tumor suppression activity.[1,4] Transfecting human mammary carcinoma cells with the maspin gene reduces tumor induction and metastasis in nude mice and in vitro invasion of basement membrane.[2] In primary breast cancer cells, maspin is downregulated and its expression is inhibited in metastasis. It has been suggested that this decrease in maspin expression may be due to the absence of transactivation through the ets and Ap1 elements in the promoter of maspin gene.[5] The expression of maspin can also be repressed by a negative hormonal response element in the prostate cells.[6] Therefore, the level of maspin expression in cancer cells may be primarily regulated by a transcriptional control. Interestingly, it has been shown recently that p53 directly upregulates the expression of maspin in breast and prostate cancer cell lines.[7] Furthermore, DNA-damaging agents and cytotoxic drugs induce endogenous maspin expression in cancer cells containing wild-type p53, but not in cells containing mutant p53.[7] Since expression of maspin inhibits the invasiveness and motility of breast and prostate tumor cells, these results suggest that maspin and p53 may cooperate in the negative regulation of tumor cell invasion and metastasis. However, no clinical data have been reported regarding possible association between the expression level of maspin and the survival rate. It is not clear whether maspin expression could provide any prognostic value for cancer patients.

The mechanism underlying the function of maspin as a tumor suppressor is not fully elucidated yet. Maspin seems to act on the cell membrane to inhibit cell motility and invasion,[8] and it has been found in many human organs at the epithelia.[3] Treatment of human breast and prostate cancer cells with recombinant maspin protein in culture inhibits invasion and motility of these cells, and this inhibition was reversible by an

Maspin, edited by Mary J.C. Hendrix. ©2002 Eurekah.com

anti-maspin antibody, suggesting that maspin itself inhibits tumor cell invasion.[8,9] A single-chain tissue plasminogen activator (tPA) specifically interacts with the maspin reactive site loop peptide and forms a stable complex with the recombinant maspin protein, suggesting that maspin may have an inhibitory effect on tPA.[10] Moreover, maspin has been recently shown to be an inhibitor of angiogenesis by using both in vitro and in vivo models.[11] The authors demonstrated that maspin acted directly on cultured vascular endothelial cells to inhibit their migration towards basic fibroblast growth factor and vascular endothelial growth factor, and thus reduced their ability to form tubular structures. Recombinant wild-type maspin protein, but not the defective mutant proteins, inhibited neovascularization in the rat cornea pocket assay, and blocked tumor growth and the associated vascularization of maspin-treated human prostate cancer cells in a xenograft mouse model.[11] However, the molecular mechanism underlying this anti-angiogenesis activity is still unknown, and it will be of interest to elucidate specific signaling pathways that cause maspin-induced apoptosis of vascular endothelial cells.

Although loss or downregulation of maspin has been implicated in tumor progression earlier, a correlation between maspin expression and survival in human cancers has just been established recently.[12] We have shown that *HER-2/neu* overexpression is significantly associated with poor survival of patients with oral squamous cell carcinoma (SCC).[13] However, the molecular pathogenesis of oral SCC is still not clear. Several reports have demonstrated a high frequency of *p53* gene mutations in head and neck SCC.[14-16] The presence of other oncogenes, such as EGFR,[17] c-*myc*,[18] and *bcl*-1[19] has also been reported in head and neck SCC. Further understanding of the molecular alterations in oral SCC will enable us to provide more accurate and useful prognostic markers and more effective treatments. Thus, we have examined the expression of maspin in oral tumor specimens and its relationship with the maspin expression and survival of patients with oral SCC. We found that decreases in maspin expression were frequent in oral SCC, and high expression of maspin was significantly associated with better survival.

Higher Maspin Expression is Associated with the Absence of Lymph Node Metastasis

We have determined the expression level of maspin in the 44 oral SCC patients' specimens by using immunohistochemical technique. A normal oral epithelial tissue taken from gingiva was used as a positive control. Of 44 patient specimens examined, 29 (66%) expressed low or intermediate levels of maspin. The tumor cells in both the low and intermediate subgroups showed much weaker staining than those of the normal tissues (positive control). The remaining 15 specimens (34%) expressed high levels of maspin and the tumor cells stained as strongly as the normal oral epithelial cells (control). The expression levels of maspin detected by immunohistochemical staining were also confirmed by Northern blot analyses.

We further examined the association of maspin with other clinicopathological features, i.e., patient age, sex, histological grade, TNM stage, and postsurgical treatment, in the 44 oral SCC patients. A significant positive association with the presence of lymph node involvement (34%, $p= 0.009$) was found in the 29 patients with low to intermediate maspin expression, but not with the other factors (Table 1). Furthermore, all of the 15 patients who expressed high levels of maspin did not show any nodal

Table 1. Clinicopathological features of oral squamous cell carcinoma and maspin immunostaining results

	Maspin expression level		
	low and intermediate[a]	high	*p* value[b]
Patients' sex			
Men	20	9	0.74
Women	9	6	
Age			
< 50 years old	11	10	0.11
≥ 50 years old	18	5	
Histological Grade			
I	21	13	0.45
II	8	2	
Tumor Size			
T_{1-2}	27	14	1.0
T_3	2	1	
Nodal Stage			
No	19	15	0.009*
Yes	10	0	
Distant Metastasis			
No	21	13	0.45
Yes	8	2	
Post-surgical Therapy[c]			
No	22	14	0.28
Yes	4	0	

[a] Patients with low and intermediate level (lower than normal level) were grouped together to perform a valid Fisher's exact test
[b] *p* value was obtained by Fisher's exact test
[c] Of the 44 patients, two received chemotherapy, and two had radiotherapy after surgery, four missing data, and the remaining 36 did not received any post surgical chemo- or radiotherapy.

involvement, suggesting that higher maspin expression is associated with the absence of lymph node metastasis. However, we did not find any association between maspin protein level and distant metastasis.

Tumor metastasis is a multistep process that includes local invasion, angiogenesis, disruption of adhesion to the neighboring cells and the extracellular matrix, adhesion to and transgression of endothelial cells to access into and out of the vascular circulation, attachment and proliferation at distant sites, adhesion to the extracellular matrix, and finally neovascularization at the distant site. Since maspin has been shown to inhibit cell invasiveness and motility,[8] our findings suggest that maspin may be involved in inhibiting the initial step in the metastasis of oral SCC.

Higher Maspin Expression is Correlated with Better Survival

Since maspin functions as a tumor suppressor, we examined whether higher maspin expression is associated with better overall survival in patients with oral SCC in our series of 44 cases. Indeed, the patients with high maspin expression had better survival rates than those who had low and intermediate levels of maspin expression (p= 0.017)

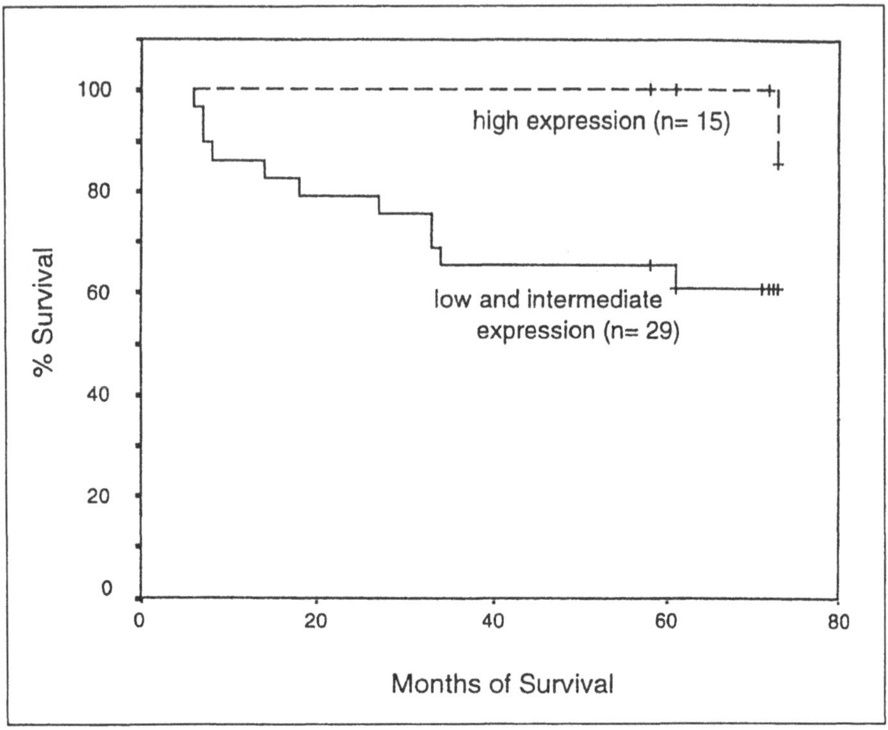

Fig. 1. Survival curves for 44 patients expressing low and intermediate maspin levels vs high maspin level (log rank test, $p = 0.017$).

(Fig. 1). These results suggest that high maspin expression in oral SCC may be a marker of favorable prognosis.

In human breast cancer, it has been reported recently that immunohistochemical staining for maspin is diagnostically useful and superior to metallothionein, S-100 protein, and alpha-smooth muscle actin, in distinguishing a radial sclerosing lesion from tubular carcinoma of the breast.[20] However, the number of patients studied in this report was relatively small and thus the potential prognostic significance of this marker in breast cancer remains to be confirmed.

Conclusions

Maspin is a tumor suppressor that has inhibitory effects on tumor invasion, metastasis, and angiogenesis. We have reviewed evidence that high maspin expression is associated with a favorable prognosis in oral SCC. The significant inverse correlation between maspin expression level and lymph node involvement suggests that loss of maspin expression may contribute to the progression of tumor metastasis. Thus, it is worthwhile to investigate a larger cohort of oral SCC and other tumor tissues to evaluate the prognostic value of maspin in human cancers.

Acknowledgements

This study was supported in part by M. D. Anderson Faculty Achievement Award (to M.C.H), and M. D. Anderson Nellie Connally Breast cancer Research Fund (to M.C.H).

References

1. Sager, R, Sheng, S, Pemberton, P et al. Maspin: A tumor suppressing serpin. Curr Topics Microbiol Immunol 1996; 213:51-64.
2. Zou Z, Anisowicz A, Hendrix MJ et al. Maspin, a serpin with tumor-suppressing activity in human mammary epithelial cells. Science 1994; 263(5146):526-529.
3. Pemberton PA, Tipton AR, Pavloff N et al. Maspin is an intracellular serpin that partitions into secretory vesicles and is present at the cell surface. J Histochem Cytochem 1997; 45(12):1697-1706.
4. Hendrix MJ. De-mystifying the mechanism(s) of maspin. Nat Med 2000; 6(4):374-376.
5. Zhang M, Maass N, Magit D et al. Transactivation through ets and Ap1 transcription sites determines the expression of the tumor-suppressing gene maspin. Cell growth Differ 1997; 8(2):179-186.
6. Zhang M, Magit D, Sager R. Expression of maspin in prostate cells is regulated by a positive ets element and a negative hormonal responsive element site recognized by androgen receptor. Proc Natl Acad Sci USA 1997; 94(11):5673-5678.
7. Zou Z, Gao C, Nagaich AK et al. p53 regulates the expression of the tumor suppressor gene maspin. J Biol Chem 2000; 275(9):6051-6054.
8. Sheng S, Carey J, Seftor EA et al. Maspin acts at the cell membrane to inhibit invasion and motility of mammary and prostatic cancer cells. Proc Natl Acad Sci USA 1996; 93(21):11669-11674.
9. Sheng, S, Pemberton, PA, Sager, R. Production, purification, and characterization of recombinant maspin proteins. 1994; J Biol Chem 269:30988-30993.
10. Sheng S, Truong B, Fredrickson D et al. Tissue-type plasminogen activator is a target of the tumor suppressor gene maspin. Proc Natl Acad Sci USA 1998; 95(2):499-504.
11. Zhang M, Volpert O, Shi YH et al. Maspin is an angiogenesis inhibitor. Nat Med 2000; 6(2):196-199.
12. Xia W, Lau YK, Hu MC et al. High tumoral maspin expression is associated with improved survival of patients with oral squamous cell carcinoma. Oncogene 2000; 19(20):2398-2403.
13. Xia W, Lau YK, Zhang HZ et al. Strong correlation between c-erbB-2 overexpression and overall survival of patients with oral squamous cell carcinoma. Clin Cancer Res 1997; 3(1):3-9.
14. Ahomadegbe JC, Barrois M, Fogel S et al. High incidence of p53 alterations (mutation, deletion, overexpression) in head and neck primary tumors and metastases; absence of correlation with clinical outcome. Frequent protein overexpression in normal epithelium and in early non-invasive lesions. Oncogene 1995; 10(6):1217-1227.
15. Somers KD, Merrick MA, Lopez ME et al. Frequent p53 mutations in head and neck cancer. Cancer Res 1992; 52(21):5997-6000.
16. Watling DL, Gown AM, Coltrera MD. Overexpression of p53 in head and neck cancer. Head Neck 1992; 14(6):437-444.
17. Yamamoto T, Kamata N, Kawano H et al. High incidence of amplification of the epidermal growth factor receptor gene in human squamous carcinoma cell lines. Cancer Res 1986; 46(1):414-416.
18. Field JK, Spandidos DA, Stell PM et al. Elevated expression of the c-myc oncoprotein correlates with poor prognosis in head and neck squamous cell carcinoma. Oncogene 1989; 4(12):1463-1468.
19. Berenson JR, Yang J, Mickel RA. Frequent amplification of the bcl-1 locus in head and neck squamous cell carcinomas. Oncogene 1989; 4(9):1111-1116.
20. Lele SM, Graves K, Gatalica Z. Immunohistochemical detection of maspin is a useful adjunct in distinguishing radial sclerosing lesion from tubular carcinoma of the breast. Appl Immunohistochem Molecul Morphol 2000; 8(1):32-36.

Index